Biological Treatment of Organic Waste in Wastewater—Towards a Circular and Bio-Based Economy

Biological Treatment of Organic Waste in Wastewater—Towards a Circular and Bio-Based Economy

Editor

Marianna Garfí

MDPI • Basel • Beijing • Wuhan • Barcelona • Belgrade • Manchester • Tokyo • Cluj • Tianjin

Editor
Marianna Garfí
Technical University of Catalonia (UPC)
Spain

Editorial Office
MDPI
St. Alban-Anlage 66
4052 Basel, Switzerland

This is a reprint of articles from the Special Issue published online in the open access journal *Water* (ISSN 2073-4441) (available at: https://www.mdpi.com/journal/water/special_issues/Circular_Bio-Based_Economy).

For citation purposes, cite each article independently as indicated on the article page online and as indicated below:

LastName, A.A.; LastName, B.B.; LastName, C.C. Article Title. *Journal Name* **Year**, *Volume Number*, Page Range.

ISBN 978-3-0365-3375-9 (Hbk)
ISBN 978-3-0365-3376-6 (PDF)

Contents

About the Editor

Marianna Garfí is an internationally recognized expert in low-cost and nature-based technologies for waste and wastewater treatment, and also in sustainability assessment (mainly life cycle assessment and multicriteria analysis). She obtained her MSc in Environmental Engineering from the Universitá di Bologna (Italy) (2005), and in 2009, she presented her PhD Thesis at the same University. After her PhD Thesis defense, she joined the Group of Environmental Engineering and Microbiology (GEMMA) of the Universitat Politècnica de Catalunya (UPC) (Spain). Currently, Dr. Marianna Garfí is leading the research line on sustainability assessment and nature-based technologies for waste and wastewater treatment at GEMMA-UPC. Her research activities aim to improve low-cost and nature-based technologies for waste and wastewater treatment (e.g., anaerobic digesters, constructed wetlands, algae-based systems) considering technical, socioeconomic, and environmental aspects.

water

MDPI

Editorial

Biological Treatment of Organic Waste in Wastewater—Towards a Circular and Bio-Based Economy

Marianna Garfí

Group of Environmental Engineering and Microbiology (GEMMA), Department of Civil and Environmental Engineering, Universitat Politècnica de Catalunya BarcelonaTech (UPC), 08034 Barcelona, Spain; Marianna.garfi@upc.edu

Due to population growth, accelerated urbanization, and economic development, the quantity of both industrial and urban wastewater generated, and its overall pollution load are increasing globally. In this context, the management of organic waste/sub-products from wastewater is an issue of great concern.

Traditionally, waste has been considered as something that is not useful and has been often neglected over the years. However, the world economic model is currently undergoing a paradigm shift from linear (waste-producing) to circular (waste-to-resources) and bio-based (using renewable biological resources) economies. Thus, there is a need to investigate innovative and cost-effective technologies and processes for the safe and environmentally friendly management of organic waste generated in wastewater treatment systems.

In this context, the biological treatment of organic waste/sub-products from both urban and industrial wastewater is a promising solution to reduce energy and the carbon footprint associated with their treatment and to shift the paradigm from waste treatment to resource recovery.

This Special Issue (SI) focuses on innovative solutions for the biological treatment of organic waste from wastewater. In particular, the research articles included in this SI are related to:

- Process mechanisms and operation, optimization, monitoring, modelling, and applications;
- Removal of pathogens and emerging pollutants;
- Reuse and circular economy;
- Resource recovery (e.g., nutrients recovery, high-value compounds) and energy valorisation (e.g., biogas);
- Life cycle assessment and carbon footprint;
- Tecno-economic assessment and social perception of waste-to-resource processes;
- Low-cost technologies;
- Policy.

Lanko et al. (2021) [1] compared the digestate quality of single-stage mesophilic and thermophilic AD and TPAD systems, in terms of the dewaterability, pathogenic safety and lower calorific value (LCV) and, based on the comparison, consider digested sludge final disposal alternatives. The results showed that TPAD system is the most beneficial in terms of organic matter degradation efficiency.

Mendieta et al. (2021) [2] analyse NCS producers' behavioural intention to use LCB by utilizing an extended technology acceptance model (TAM). This study's findings contribute to research on the TAM and provide a better understanding of the factors influencing NCS producers' behavioural intention to use low-cost digesters.

Lanko et al. (2020) [3] investigated the environmental impact of the anaerobic digestion (AD) of sewage sludge within an activated sludge wastewater treatment plant (WWTP). Three alternative AD systems (mesophilic, thermophilic, and temperature-phased anaerobic digestion (TPAD)) were compared to determine which system may have the best environmental performance. The results showed that the best AD alternative was

Citation: Garfí, M. Biological Treatment of Organic Waste in Wastewater—Towards a Circular and Bio-Based Economy. *Water* **2022**, *14*, 360. https://doi.org/10.3390/w14030360

Received: 20 January 2022
Accepted: 21 January 2022
Published: 26 January 2022

thermophilic concerning all environmental impact categories, besides climate change and human toxicity.

Kassab et al. (2020) [4] proposed a potential approach for enhanced energy generation from anaerobic digestion; iron-based conductive nanoparticles have been proposed to enhance the methane production yield and rate. The results have shown that supplementing anaerobic batches with NZVIs has an insignificant impact, most probably due to the agglomeration of NZVI particles and, consequently, the reduction in available surface area, making the applied doses insufficient for measurable effect.

Zhang et al. (2020) [5] provided a reference for the application of heterotrophic nitrification-aerobic denitrification in actual wastewater treatment. From the results, the synthetic microbial community was able to simultaneously perform heterotrophic nitrification-aerobic denitrification indicating great potential for full-scale applications.

In conclusion, this SI provided new ways to valorise organic waste from wastewater and describe novel processes, as well as the environmental and social benefits in the frame of the Sustainable Development Goals.

Funding: This research was funded by the Government of Catalonia (Consolidated Research Group 2017 SGR 1029), and the Spanish Ministry of Economy and Competitiveness (RYC-2016 20059).

Institutional Review Board Statement: Not applicable.

Informed Consent Statement: Not applicable.

Data Availability Statement: Not applicable.

Acknowledgments: Marianna Garfí is grateful to the Government of Catalonia (Consolidated Research Group 2017 SGR 1029), and the Spanish Ministry of Economy and Competitiveness (RYC-2016 20059).

Conflicts of Interest: The author declares no conflict of interest.

References

1. Lanko, L.; Hejnic, J.; Říhová-Ambrožová, J.; Ferrer, I.; Jenicek, P. Digested Sludge Quality in Mesophilic, Thermophilic and Temperature-Phased Anaerobic Digestion Systems. *Water* **2021**, *13*, 2839. [CrossRef]
2. Mendieta, M.; Castro, L.; Vera, V.; Rodríguez, J.; Escalante, H. Toward the Adoption of Anaerobic Digestion Technology through Low-Cost Biodigesters: A Case Study of Non-Centrifugal Cane Sugar Producers in Colombia. *Water* **2021**, *13*, 2566. [CrossRef]
3. Lanko, I.; Flores, L.; Garfi, M.; Todt, V.; Posada, J.A.; Jenicek, P.; Ferrer, I. Life Cycle Assessment of the Mesophilic, Thermophilic and Temperature Phased Anaerobic Digestion of Sewage Sludge. *Water* **2020**, *12*, 3140. [CrossRef]
4. Kassab, G.; Khater, D.; Odeh, F.; Shatanawi, K.; Halalsheh, M.; Arafah, M.; van Lier, J. Impact of Nanoscale Magnetite and Zero Valent Iron on the Batch-Wise Anaerobic Co-Digestion of Food Waste and Waste-Activated Sludge. *Water* **2020**, *12*, 1283. [CrossRef]
5. Zhang, Q.; Yang, P.; Liu, L.; Liu, Z. Formulation and Characterization of a Heterotrophic Nitrification-Aerobic Denitrification Synthetic Microbial Community and its Application to Livestock Wastewater Treatment. *Water* **2020**, *12*, 218. [CrossRef]

water

MDPI

Article

Digested Sludge Quality in Mesophilic, Thermophilic and Temperature-Phased Anaerobic Digestion Systems

Iryna Lanko [1,2,*], Jakub Hejnic [1], Jana Říhová-Ambrožová [1], Ivet Ferrer [2] and Pavel Jenicek [1]

1 Department of Water Technology and Environmental Engineering, University of Chemistry and Technology, Technicka, 5, 166 28 Prague, Czech Republic; jakub.hejnic@gmail.com (J.H.); ambrozoj@vscht.cz (J.Ř.-A.); jenicekp@vscht.cz (P.J.)

2 GEMMA—Group of Environmental Engineering and Microbiology, Department of Civil and Environmental Engineering, Universitat Politècnica de Catalunya·BarcelonaTech, c/Jordi Girona 1–3, Building D1, 08034 Barcelona, Spain; ivet.ferrer@upc.edu

* Correspondence: irynalanko@gmail.com

Abstract: Anaerobic digestion (AD) technology is commonly used to treat sewage sludge from activated sludge systems, meanwhile alleviating the energy demand (and costs) for wastewater treatment. Most often, anaerobic digestion is run in single-stage systems under mesophilic conditions, as this temperature regime is considered to be more stable than the thermophilic one. However, it is known that thermophilic conditions are advantageous over mesophilic ones in terms of methane production and digestate hygienisation, while it is unclear which one is better concerning the digestate dewaterability. Temperature-phased anaerobic digestion (TPAD) is a double-stage AD process that combines the above-mentioned temperature regimes, by operating a thermophilic digester followed by a mesophilic one. The aim of this study is to compare the digestate quality of single-stage mesophilic and thermophilic AD and TPAD systems, in terms of the dewaterability, pathogenic safety and lower calorific value (LCV) and, based on the comparison, consider digested sludge final disposal alternatives. The research is conducted in lab-scale reactors treating waste-activated sludge. The dewaterability is tested by two methods, namely, centrifugation and mechanical pressing. The experimental results show that the TPAD system is the most beneficial in terms of organic matter degradation efficiency (32.4% against 27.2 for TAD and 26.0 for MAD), producing a digestate with a high dewaterability (8.1–9.8% worse than for TAD and 6.2–12.0% better than for MAD) and pathogenic safety (coliforms and *Escherichia coli* were not detected, and *Clostridium perfringens* were counted up to 4.8–4.9 $\times$ 10^3, when for TAD it was only 1.4–2.5 $\times$ 10^3, and for MAD it was 1.3–1.8 $\times$ 10^4), with the lowest LCV (19.2% against 15.4% and 15.8% under thermophilic and mesophilic conditions, respectively). Regarding the final disposal, the digested sludge after TAD can be applied directly in agriculture; after TPAD, it can be used as a fertilizer only in the case where the fermenter HRT assures the pathogenic safety. The MAD digestate is the best for being used as a fuel preserving a higher portion of organic matter, not transforming into biogas during AD.

Keywords: mesophilic; thermophilic; temperature-phased anaerobic digestion (TPAD); dewaterability; sludge quality; sludge valorisation

Citation: Lanko, I.; Hejnic, J.; Říhová-Ambrožová, J.; Ferrer, I.; Jenicek, P. Digested Sludge Quality in Mesophilic, Thermophilic and Temperature-Phased Anaerobic Digestion Systems. *Water* **2021**, *13*, 2839. https://doi.org/10.3390/w13202839

Academic Editor: Alicia Ronda Gálvez

Received: 30 August 2021
Accepted: 7 October 2021
Published: 12 October 2021

1. Introduction

Nowadays, sustainable sewage sludge management shifts to introduce the implementation of a resource recovery approach rather than only dispose produced sludge. It turns WWTPs into water resource recovery facilities (WRRF) [1,2]. Hence, the sludge is converted into energy, nutrients, and other valuable substances (metals, specific organic substances). All of the above mentioned can be reused in different spheres of our life, including agriculture (fertilizers), various industries (biopolymers, fuels) and communal services (heat) [2–4]. By this, lower emissions of pollutants to the environment are reached [5]. Consequently, a better environmental protection level is achieved.

The recovery of resources and final reuse may cover around 30% of the costs for sewage sludge handling [6], which is an important amount, given that the sewage sludge handling usually takes up to 50% of the wastewater treatment expenses [7].

Sewage sludge from activated sludge WWTP comprises the so-called primary sludge, produced during the primary wastewater treatment in sedimentation tanks; secondary sludge is produced during the secondary wastewater treatment in biological reactors as a result of microbial growth. Normally, the ratio of the produced sludge types may vary from 40 to 70% of primary to secondary sludge [8,9]. At small and medium WWTPs (<50,000 PE), the primary sedimentation step may be absent, having only secondary sludge production. Secondary sludge is also known as waste-activated sludge (WAS) [10].

Anaerobic digestion (AD) is one of the most widespread and favourable means of sewage sludge handling at medium to large WWTPs using the activated sludge system. AD consists of the biodegradation of organic matter under anaerobic conditions, leading to the production of biogas (mostly composed of methane) and a stabilised digestate [11]. The AD process has four steps, namely, rate-limiting hydrolysis, acidogenesis, acetogenesis and methanogenesis [12]. Each of these steps is performed by a specific type of bacteria or archaea. During sludge digestion, hydrolysis is the rate-limiting process; therefore, innovative technologies such as the TPAD are trying to improve this step. AD may be carried out under different temperature conditions, namely, psychrophilic (0–20 °C), mesophilic (35–40 °C) and thermophilic (50–60 °C), by using single-stage or double-stage processes [11,12]. At the single-stage AD process, all four steps of AD take place in the same reactor simultaneously. In this case, all types of bacteria and/or archaea (under thermophilic conditions, methanogenic microorganisms are represented only by archaea) have to co-survive in a restricted range of pH values (±1.0 pH) [13], controlled by an organic loading rate (OLR) and hydraulic retention (HRT). The pH balance is an important monitored factor that helps to avoid the inhibition of methanogens known as slow growers under an increased OLR and/or shortened HRT. It is also important to mention that, along with increasing the AD temperature, when switching from mesophilic to thermophilic or even hypothermophilic conditions, the pH balance starts to be the more crucial aspect that can lead not only to a limited methane production, but also to a complete digester failure [14]. Due to this, there are a lot of studies conducted on different pre-treatment applications in order to promote methane production [15,16]. The so-called temperature-phased anaerobic digestion (TPAD) consists of two stages which are carried out in two anaerobic reactors implemented in series, a thermophilic followed by a mesophilic one [17,18]. TPAD systems combine the advantages of single-stage mesophilic and thermophilic systems by splitting the different types of microorganisms physically: the first two AD steps take place in the first reactor and the other two AD steps happen in the second reactor. This allows to manipulate the conditions of the rate-limiting hydrolysis by increasing the temperature and/or increasing the OLR/shortening HRT, in a much broader way excluding the direct negative influence on the methanogens located in the second digester, which promotes a higher organic matter degradation rate and, consequently, a higher methane production [19].

With regard to the environmental impacts, AD systems with the higher efficiency of organic matter degradation are more environmentally friendly. In terms of the whole WWTP, the life cycle assessment analysis showed the lowest burden on the environment from TPAD and, within the sludge line, TAD and TPAD were more beneficial than the more stable in operation MAD [2].

There are plenty of full-scale references of single-stage anaerobic digestion systems in Europe under both mesophilic and thermophilic conditions [20]. At the same time, there is no such variety of temperature-phased anaerobic digestion system examples [21], even though its beneficial performance at lab-scale [22] and full-scale [7] has already been proved. There have been several studies conducted in recent decades on TPAD efficiency over single-stage mesophilic and thermophilic reactors [21,23,24]. To the best of our knowledge, the performance of two-stage systems outcompetes mesophilic and thermophilic single-stage systems in both organic matter degradation and methane production [24–26]. However, "to close the loop" of digestion efficiency, a digestate quality assessment is needed. In particular, digestate dewaterability is relevant in order to reduce the digestate volume and management costs, while hygienisation is an important issue upon the land application of the digestate.

Dewaterability is a complex quality parameter of sewage sludge describing the ability of sludge flocs to "lose" water, which is entrapped inside. According to the difficulty of its removal, all water that is present in the sewage sludge can be divided into three types: free water which is unaffected by solid particles, interstitial water which is physically trapped inside the space between particles, and surface water which is adsorbed onto the surface of solid particles [27–29].

There are a lot of methods for sewage sludge dewaterability characterization [30] such as the capillary suction time, filterability testing, sludge centrifugation and sludge pressing. However, it is challenging to find a good correlation between lab-scale results and full-scale dewatering efficiency. A good correlation between lab-scale and full-scale dewatering efficiency plays an important role when conducting the trials of dewatering mechanisms, choosing a flocculant and its proper dosage. The above-mentioned processes are costly and depend on the type of dewatering equipment [31], and at lab-scale its performance could be cheaper and faster with the same extent of reliability. Hence, this research focuses on two methods of mechanical dewatering that are, in principle, similar to the dewatering processes used in full-scale WWTP, namely, centrifugation and mechanical pressing, and the calculation of the universal parameter of dewaterability obtained by [31].

The lower calorific value (LCV) is an important indicator of the efficient energy recovery from the digested sludge by incineration, pyrolysis, gasification, etc. The LCV is determined by the original sludge composition, degradation efficiency and dewaterability. Regarding the LCV of digested sludge, the efficiency of dewatering plays a major role, as poor dewaterability means a large amount of water in the sludge and this results in a low (often negative) LCV, because the energy value of organic matter in the sludge is lower than the energy needed to evaporate the water present. Hence, the LCV provides additional information on the quality of the digested sludge which helps to select the optimal final disposal solution from both an economic and environmental point of view [32].

Finally, pathogen removal is one of the crucial parameters for a safe treated sludge reuse in agricultural land. The legislation strictly defines which pathogen removal extent should be reached for each type of sewage sludge's final disposal, especially in the case of use as a fertiliser in agriculture [33].

The aims of the study are to provide a comprehensive comparison on the single-stage mesophilic and thermophilic AD and TPAD systems in terms of the process performance (organic matter removal and methane production) and digestate quality (dewaterability, pathogenic safety and energy value expressed as a lower calorific value) and, based on the obtained data, suggest the best alternatives for final digested sludge disposal.

2. Materials and Methods

2.1. Experimental Set-Up

The experimental laboratory set-up consisted of three AD systems: single-stage thermophilic, single-stage mesophilic and double-stage TPAD (Figure 1).

The reactors were composed of thermoresistant plastic with standing temperatures of up to 60 °C. The mesophilic, thermophilic and second stage of the TPAD reactors had a working volume of 8.45 L, while the first stage of the TPAD (fermenter) had a working volume of 1.45 L, all of them with a headspace volume of around 1.0 L. The feeding and wasting processes were automated and governed by LabVIEW 2012 software version 12.0 (32-bit) ran on embedded controller cRIO 9074 (both, the software and controller from National Instruments, Prague, Czech Republic). There were three programmed cased drive peristaltic tube pumps (Verderflex Vantage 3000 P R3I EU, Verder s.r.o., Prague, Czech Republic): the feeding pump; the TPAD pump that transferred pre-digested sludge from the first stage to the second stage of TPAD; the wasting pump (Figure 1). All pumps were calibrated at least once per month, as all AD digestates and WAS used as a substrate were non-Newtonian, viscous fluids.

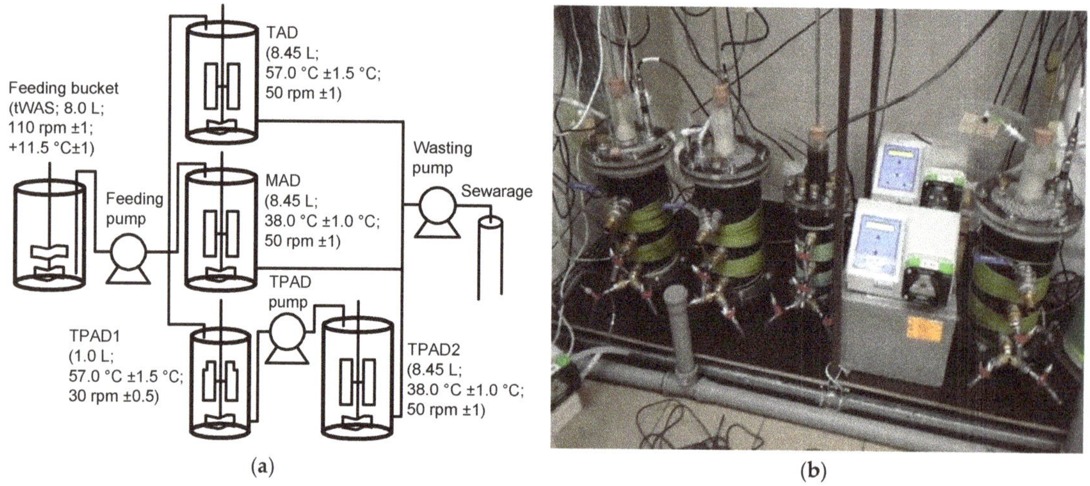

Figure 1. Scheme (**a**) and picture (**b**) of lab-scale experimental set-up.

At first, the digestate was withdrawn from TPAD2; then, TPAD pump added pre-digested sludge from TPAD1 to TPAD2; then, the fermenter was fed by running the feeding pump. After that, the digestate was withdrawn from the single-stage mesophilic reactor and, immediately after that, it was fed. Finally, some digestate was taken from the thermophilic reactor and the same amount of substrate was added back to substitute for the withdrawn volume. The whole cycle took around thirty minutes and happened once per twenty-four hours (semi-continuous model of reactor feeding) at the same time.

The gas meter RITTER MilliGascounter (RITTER Apparatebau GmbH and Co, Bochum, Germany) were used to estimate the biogas flow.

All data were monitored online and logged in.

The whole period of experiments was divided into two phases: Phase A and Phase B. Phase A lasted for 5 months (HRT = 19.0 days; ORL = 2.24–2.25 kg VS·m^{-3}·day^{-1}) and Phase B for 3 months (HRT = 13.5 days; ORL. = 3.58–3.62 kg VS·m^{-3}·day^{-1}). The AD performance of single- and double-stage systems was evaluated in terms of organic matter degradation, methane production and digestate quality. To do so, the two above-mentioned sets of experiments were conducted, with the following operational parameters (Table 1):

Table 1. Anaerobic digestion operational parameters.

Type of Reactor	Abbreviation	Phase A, Lasting 5 Months		Phase B, Lasting 3 Months		Phase A and Phase B
		HRT, Days	Temperature Range, °C	HRT, Days	Temperature Range, °C	Mixing Speed, rpm **
Single-stage, thermophilic	TAD	19.0	57 ± 1.5 °C	13.5	57 ± 1.0 °C *	50 ± 1
Single-stage, mesophilic	MAD	19.0	38 ± 1.5 °C	13.5	38 ± 1.0 °C *	50 ± 1
Double-stage, thermophilic, the first stage	TPAD1	2.0	57 ± 1.5 °C	2.0	57 ± 1.0 °C *	30 ± 1

Table 1. *Cont.*

Type of Reactor	Abbreviation	Phase A, Lasting 5 Months		Phase B, Lasting 3 Months		Phase A and Phase B
Double-stage, mesophilic, the second stage	TPAD2	17.0	38 ± 1.5 °C	11.5	38 ± 1.0 °C *	50 ± 1

* Footnote 1. All digesters at Phase B were insulated to decrease temperature fluctuation. ** Footnote 2. All digesters were continuously mixed at the fixed speed. Footnote 3. TAD—thermophilic anaerobic digestion; MAD—mesophilic anaerobic digestion; TPAD—temperature-phased anaerobic digestion; TPAD1—the first stage of TPAD; TPAD2—the second stage of TPAD; HRT—hydraulic retention time.

The reactors were inoculated with digested sludge from the full-scale anaerobic digesters at Czech municipal wastewater treatment plants and thickened waste-activated sludge (WAS) was used as a substrate. The substrate was kept in the fridge under 11.5 ± 1.0 °C, continuously mixed at 110 ± 2 rpm. The sludge samples were characterized in terms of total suspended solids (TSS), volatile suspended solids (VSS), total and soluble chemical oxygen demand (tCOD and sCOD), pH.

The start-up period lasted fifty days (about threefold of HRT). To monitor the reactors performance, the following parameters were analysed regularly: pH (online), temperature (online), biogas volume (every day), biogas composition (three times per week), volatile fatty acid (VFA), VSS and TSS contents (once per week). The digestate quality was evaluated by measuring the dewaterability, hygienisation and LCV, as described in the following sections.

2.2. Digestate Dewaterability

The digestate dewaterability, was evaluated by two methods: (1) separation via centrifugation; (2) filtration and compression via mechanical pressing.

2.2.1. Centrifugation

The principle of this method is measuring the sludge cake concentration after centrifugation. For this method, all samples were centrifuged in Sigma 3–16 P (SIGMA Laborzentrifugen GmbH, Osterode am Harz, Germany) at 13,083 rpm for 10 min, and the weight of the separated fugate was measured. The higher the weight, the better the dewaterability of the sludge.

Digestate samples were centrifuged and, then, the weight of the separated fugate and sludge cake were measured. Afterwards, the concentration of TS in the sludge cake was calculated as a ratio between the amount of TS in the sample and the weight of the separated sludge cake.

The dewaterability of the digestate was calculated by the dewaterability coefficient (%), calculated from (Equation (1)):

$$\frac{Wdry\ matter}{Wsludge\ cake} \times 100\% \tag{1}$$

where $W_{dry\ matter}$ is the weight of dry matter in the centrifuged sample (g); $W_{sludge\ cake}$ is the weight of sludge cake after centrifugation (g).

It is important to note that separation by centrifugation characterized the quality of the original digested sludge without flocculant addition.

2.2.2. Mechanical Pressing

This method was carried out using a mini-press Mareco MMP-3/2 (Amfitech Friesland BV, Joure, The Netherlands) (Figure 2).

Figure 2. Laboratory mini-press (Mareco MMP-3/2).

The experimental procedure was as follows: Initially, the TS concentration of digestate samples from each reactor was determined. Following, 1 L of concentrated polymer SUPERFLOC C-494HMW (Kemifloc a.s., Prerov, Czech Republic) solution (5.0 g/L) was prepared and used within 4–8 h. Tap water was used to prepare the suspension. Then, it was mixed thoroughly at 1000 rpm by a blade impeller until no flocs were observed. The corresponding volume (18–23 mL) of flocculant stock solution was then added to 50 mL of digestate, which was defined in advance based on the TS concentration measurement and for each type of digestate. The digestate sample with flocculant dosage was then mixed at 700 rpm for 3 min (Figure 3).

(a) (b) (c)

Figure 3. Digestate dewatering by mechanical pressing: (**a**) digestate with flocculant dosage; (**b**) mixing at 700 rpm for 3 min; (**c**) dewatered digestate.

The pressing was performed under 5 bar and lasted 1000 s (a full cycle of mini-press, Mareco). The supernatant (Figure 3a) was weighted on the calibrated analytical balance Acculab ALC-3100.2. The quality of supernatant in terms of its cleanness was assessed every time visually. The TS concentration of the sludge cake produced by sludge pressing was determined.

2.3. Elemental Analysis and Lower Calorific Value

The elemental analysis (EA) of the digestate was performed in order to calculate two parameters—the lower calorific value and universal factor to describe AD substrates—which were used to estimate the sludge cake TSS concentration in full-scale AD [31], the so-called $C/N \times ash$ parameter (Equation (2)):

$$5.53 \times \frac{C}{N} \times ash + 7.14 \tag{2}$$

where C/N is the ratio between C and N content in the digestate; *ash* is the mass fraction (1-VSS/TSS) (the empirical values obtained experimentally were 5.53 and 7.14 for VSS and TSS, respectively).

For the EA, an integrated sample was collected for each type of digestate over a period of 4 days in a row, and dried at 105 °C. EA was performed in triplicate every other week with the Elementar vario EL Cube (Elementar Analysensysteme GmbH, Langenselbold, Germany).

The lower calorific value was calculated based on the average data on ash content from the EA of digestate samples. Thus, the $LCVsludge$ ($kJ \cdot kg^{-1}$) was calculated according to (Equation (3)) [34]:

$$LCVsludge = 4.18 \times (94.19 \times C - 0.5501 - 52.14 \times H) \tag{3}$$

where 4.18, 92.19, 0.5501, 52.14 are empirical coefficients calculated on the basis of experimental data; C is the carbon weight fraction (%); H is the hydrogen weight fraction, (%).

2.4. Pathogenic Bacteria Indicators

Digestate hygienisation was evaluated by assessing the pathogenic bacteria indicators. Firstly, digestate samples were pre-treated as follows: 1 g of a digested sludge sample was diluted in 9 mL of physiological solution (9 g of NaCl in 1 L of distilled water and then sterilised). Then, it was diluted to 10^{-2} and 10^{-3}. To measure the total counts of bacteria, as indicators of organotrophic and faecal contamination, the following microorganisms were chosen: culturable aerobic microorganisms cultivated at 22 °C and 36 °C [35], total coliforms and *E. coli* [36,37] and Clostridium perfringens [38]. The cultivation procedure for microorganisms cultivated at 22 °C and 36 °C was as described below: 1 mL of the pre-treated and diluted digested sludge sample was added to a Petri dish; then, the sterilised growth medium [35] was poured into the Petri dish. The procedure of the faecal contamination indicators cultivation was slightly different: 0.2 mL of the pre-treated and diluted digested sludge sample was placed directly on the surface of the sterilised growth medium [36–38] placed in a Petri dish earlier.

2.5. Analytical Methods

2.5.1. Biogas Production and Composition

Biogas production was measured by the Ritter MilliGascounter "MGC-1 V3.4 PMMA" (Q_{min} = 1 mL/h; Q_{max} = 1 L/h; P_{max}: 5.0 mbar; preciseness: ±3%) from RITTER Apparatebau GmbH and Co, Bochum, Germany. The MilliGascounters were filled with the HCl 1.8% solution at the liquid phase to avoid any dissolving and outgassing processes (mainly, this relates to the presence of CO_2) to the greatest possible extent.

The biogas composition was assessed using the gas chromatograph (GC) Shimadzu GC-2014 (Shimadzu Europa, Duisburg, F.R. Germany) with a thermal conductivity detector (temperature 185 °C) and injection via on-column with packed column (packed by HayeSep D 100/120 mash; oven: isotherm 130 °C, flow 30 mL/min; carrier gas—Helium). A total of 1.0 mL of biogas produced was withdrawn with a tight syringe, and introduced into the column, which evaluated the gaseous composition. The percentage of carbon dioxide, methane and nitrogen was detected in each sample. Hydrogen content was monitored using GC 8000 Top Gas Chromatograph by CE Instruments Ltd., Hindley

Green, UK, with a thermal conductivity detector (temperature 185 °C) and injection via on-column with packed column (packed by HayeSep D 100/120 mash; oven: isotherm 100 °C, flow 30 mL/min; carrier gas—Helium). The specific methane production *Qsp.methane* (L/gCOD$_{added}$) was calculated in the following way (Equation (4)):

$$Qsp.methane = \frac{Qmethane}{(Wdose \times CODsubstrate)} \quad (4)$$

where $Q_{methane}$—daily methane production, L/day; W_{dose}—substrate volume added, L/day; $COD_{substrate}$—COD concentration in the substrate, gCOD$_{added}$/L.

2.5.2. Suspended Solids

The solid analysis of the sludge formed the basic characterization of the sample. The test determined the content of Total Solids (TS), Volatile Solids (VS), Dissolved Solids (DS), Fixed Solids (FS), Total Suspended Solids (TSS) and Volatile Suspended Solids (VSS). In order to determine the solid content of the sludge, the procedures described by Standard Methods, APHA [39] were used.

2.5.3. Chemical Oxygen Demand

All samples were analysed accordingly to the Standard Methods [39]. To determine the total COD (tCOD), the samples were usually diluted, so that measured COD values fell within the detection limits of the spectrophotometer Hach Lange DRB-3900 (Hach, Prague, Czech Republic) set at 600 nm wavelength. For boiling the samples, an incubating mineralizer Hach Lange DRB-200 (Hach, Prague, Czech Republic) was used. All samples were measured in triplicates.

2.5.4. Temperature, pH and VFA Measurement

The monitored temperature, as well as pH of the media, were measured online by means of Polilyte Plus H Arc 225 from Hamilton Bonaduz AG, Rapperswil-Jona, Switzerland. All four probes were connected to the computer and LabView 2012 software to be able to log the data online.

The VFAs were measured weekly, employing GC Shimadzu GC-2010 (Europa, Duisburg, F.R. Germany) with a flame ionization detector and capillary column CP-Vax58 of 25 m length and 0.25 mm inner diameter (HPST s.r.o., Prague, Czech Republic). The oven program was the following: 70 °C with a rate of 15 °C/min to 134 °C and isotherm for 1 min. Total time of analysis was 5.27 min. Injection temperature was 270 °C at the split mode. Detector temperature was 300 °C.

The samples were prepared by centrifuging the digestate for 10 min at 13,083 rpm in the centrifuge Sigma 3–16 P (SIGMA Laborzentrifugen GmbH, Osterode am Harz, Germany), filtered through a filter ACRODISC PSF (Filter Concept s.r.o., Ostrava, Czech Republic) with a 0.45 μm diameter pore size and diluted ten times before the measurement.

The VFA concentration *QVFA* (g/gCOD$_{added}$) was calculated in the following way (according to Equation (5)):

$$QVFA = \frac{CVFA \times Vreactor}{Wdose \times CODsubstrate} \quad (5)$$

where C_{VFA}—daily VFA concentration, g/L; $V_{reactor}$—working volume of the reactor, L; W_{dose}—substrate volume added, L/day; $COD_{substrate}$—COD concentration in the substrate, gCOD$_{added}$/L.

2.6. Statistics

For performing the statistical analysis, statistical technique ANOVA (Analysis of Variance) was used.

A one-way ANOVA technique was applied. That meant that only one independent variable—the temperature of AD process—was used. Statistical verification of significance was performed at significance level $a = 0.05$. For statistically significant results, the further Scheffé's method was applied.

The Scheffé's method was used for the multiple comparison of the average values (or contrasts). The estimation of each contrast for three procedures was defined as follows (according to Equation (6)):

$$\hat{\psi}_{i,j} = \overline{x}_i - \overline{x}_j \quad (6)$$

where $i \neq j$ and were equal, from 1 to 3, to the number of contrasts.

The Scheffé's test is the most conservative procedure as it provides the narrowest confidence interval. The confidence interval within Scheffé's test is defined as (Equation (7)):

$$\hat{\psi}_{i,j} \pm \sqrt{(I-1) \times s \times F \times (\frac{1}{r_i} + \frac{1}{r_j})} \quad (7)$$

where $\hat{\psi}_{i,j}$ is the *i-*, *j-* contrast, *I* is a number of parameter levels (in this case, $I = 3$), r_i, r_j is a number of repetition in *i-*, *j-* levels; *s*—the residual standard deviation (from ANOVA), *F* is the critical *F*-value for *(1-a)* and *((I-1)*; *(N-I))* degrees of freedom, *N* is the total number of experiments in ANOVA table.

If the confidence interval for *i-*, *j-* contrast contained zero value, the contrast was non-significant.

3. Results and Discussion

3.1. Anaerobic Digestion Performance

The organic matter degradation efficiency was one of the fundamental parameters of the AD process. As the substrate characteristics changed during the digester operation, the average values of VS and VSS, and removal efficiency, were calculated separately for both experimental periods: Phase A with an HRT of 19 days and Phase B with an HRT of 13.5 days (Table 2).

Table 2. Organic matter removal in the mesophilic, thermophilic and temperature-phased anaerobic digestion systems.

Phase	Parameter	TAD	MAD	TPAD1	TPAD2	Substrate
Phase A, at HRT = 19.0 days	Digestate VS, g/L	30.8 ± 1.7	31.3 ± 2.6	35.4 ± 4.1	28.6 ± 1.0	42.3 ± 4.1
	VS removal, %	27.2	26.0	16.3	32.4	-
	Digestate VSS, g/L	21.0 ± 1.8	25.2 ± 2.6	28.0 ± 3.0	22.2 ± 1.7	37.5 ± 3.4
	VSS removal, %	44.0	32.8	25.3	40.8	-
Phase B, at HRT = 13.5 days	Digestate VS, g/L	32.6 ± 0.9	32.0 ± 0.7	35.6 ± 1.0	30.1 ± 0.5	46.5 ± 0.6
	VS removal, %	22.9	24.3	15.8	28.8	-
	Digestate VSS, g/L	21.6 ± 1.2	23.7 ± 0.7	27.6 ± 0.6	20.8 ± 0.8	42.2 ± 2.1
	VSS removal, %	42.9	36.8	26.4	44.5	-

Footnote 1. VS and VSS removal in TPAD2 column express the total efficiency of the TPAD process. Footnote2. VS—volatile solids; VSS—volatile suspended solids; TAD—thermophilic anaerobic digestion; MAD—mesophilic anaerobic digestion; TPAD—temperature-phased anaerobic digestion; TPAD1—the first stage of TPAD; TPAD2—the second stage of TPAD; HRT—hydraulic retention time.

The achieved VS removal efficiency (23–32%) was relatively low, which reflected the fact that only thickened waste-activated sludge was used as a substrate, and that the systems were operated at a relatively high organic loading rate of 2.24–2.25 kg·m^{-3}·day^{-1} (Phase A) and 3.58–3.62 kg·m^{-3}·d^{-1} (Phase B) as a result of the relatively short HRT (Table 1). Similar VS removal rates (30–40%) were measured by [40] for WAS as a substrate. Oppositely, [22] registered the additional 8% of VS removal at TPAD compared to the conventional MAD. The results showed that the VS removal efficiency decreased by only 2–5% after changing from 19 days (Phase A) to 13.5 days (Phase B) of HRT: 4.3% for TAD, 1.7% for MAD and 4.1% for TPAD. In terms of VSS, there was a slight removal rate increase of 1% for TAD, which was negligible as the standard deviation was around the same value, a bigger removal rate increase of 4% for MAD and 4.8% for TPAD. This meant that shortening the HRT reduced the degradation efficiency of all AD systems. However, the acceptable efficiency was still achieved even at a significantly shortened HRT, especially in TPAD. The authors of [41] also stated higher efficiencies for the organic matter removal rate (30%) and methane production (26–60%) at TPAD than at any single-stage AD with the same HRT.

The operation of the TAD at a short retention time was the least stable, which resulted in a poor VS degradation efficiency and the accumulation of VFA (Table 3).

Table 3. The VFA concentration.

Phases	VFAs, mgCOD/L			
	TAD	MAD	TPAD1	TPAD2
Phase A, HRT = 19.0 days	3858.3 ± 973.1	519.1 ± 184.5	7451.8 ± 1777.7	522.4 ± 145.4
Phase B, HRT = 13.5 days	4821.1 ± 195.0	328.9 ± 43.2	7809.7 ± 534.0	366.8 ± 17.9

Footnote. VFA—volatile fatty acids; TAD—thermophilic anaerobic digestion; MAD—mesophilic anaerobic digestion; TPAD—temperature-phased anaerobic digestion; TPAD1—the first stage of TPAD; TPAD2—the second stage of TPAD; HRT—hydraulic retention time.

Table 3 shows the average VFA concentrations in the various digesters. The highest concentration of VFA was found in TPAD1, where acidogenesis was the aim. The high concentration of VFA in TAD indicated a lower stability of the thermophilic process under the tested conditions for both HRTs. In contrast, both mesophilic digesters (MAD and TPAD2) showed very low VFA concentrations and a stable performance at both HRTs.

The results of methane production (Table 4) corresponded well with the VS degradation efficiency (Table 2).

Table 4. Specific methane production.

Phase	Methane Production, mL/g COD_{added}				
	TAD	MAD	TPAD1	TPAD2	TPAD
Phase A, at HRT = 19 days	169.4 ± 9.2	156.0 ± 9.1	46.1 ± 4.0	186.9 ± 10.7	233.1 ± 12.0
Phase B, at HRT = 13.5 days	116.5 ± 12.0	133.9 ± 26.4	40.2 ± 8.1	132.0 ± 9.8	172.3 ± 11.6

Footnote. COD—chemical oxygen demand; HRT—hydraulic retention time; TAD—thermophilic anaerobic digestion; MAD—mesophilic anaerobic digestion; TPAD—temperature-phased anaerobic digestion; TPAD1—the first stage of TPAD; TPAD2—the second stage of TPAD; HRT—hydraulic retention time.

The double-stage TPAD system achieved the highest specific methane production in both periods: 233 mL/g COD added vs. 170 mL/g COD added for the TAD and 156 mL/g COD added for the MAD (Phase A) and 172 mL/g COD added vs. 116.5 mL/g COD added for the TAD and 134 mL/g COD added for the MAD (Phase B). Indeed, the TPAD system reached comparable results with an HRT of 13.5 days (172 mL/g COD added) to TAD and MAD with an HRT of 19 days (170 and 156 mL/g COD, respectively). According to the statistical analysis performed, the difference in methane production at both Phases was statistically significant for all AD systems. The correlations were considered statistically significant at a 95% confidence interval ($a < 0.05$). The authors of [42] also proved that TPAD showed a better performance in terms of methane production of up to 20% in comparison to the single-stage MAD.

The double-stage TPAD system in principle separated the AD stages: hydrolysis and acidogenesis took place in the 1st stage, while acetogenesis and methanogenesis occurred in the 2nd stage [43]. Therefore, the second stage of the TPAD (TAPD2) was expected to have the highest methane content in biogas (Table 5).

Table 5. Biogas composition.

Phase	Biogas Content, %							
	TAD		MAD		TPAD1		TPAD2	
	CH_4	CO_2	CH_4	CO_2	CH_4	CO_2	CH_4	CO_2
Phase A, at HRT = 19 days	61.7 ± 4.8	34.7 ± 5.0	66.1 ± 1.8	30.2 ± 3.6	58.9 ± 12.8	33.2 ± 8.4	70.9 ± 2.7	24.7 ± 2.7
Phase B, at HRT = 13.5 days	61.7 ± 1.7	33.5 ± 2.3	64.2 ± 2.1	29.9 ± 1.6	53.4 ± 4.7	39.8 ± 5.0	71.4 ± 1.3	23.8 ± 0.9

Footnote. TAD—thermophilic anaerobic digestion; MAD—mesophilic anaerobic digestion; TPAD—temperature-phased anaerobic digestion; TPAD1—the first stage of TPAD; TPAD2—the second stage of TPAD; HRT—hydraulic retention time.

Moreover, the methane content in TPAD1 was expected to be much lower because of the very short retention time (2.0 days). According to the literature, the generation time of methanogens may vary in a broad range of 0.1–12.4 days [32]. In this case, the presence of methanogens can be explained by the production of a biofilm on the digester walls and the mixing device. To our best knowledge, a much higher retention time of the biofilm in comparison with the suspended biomass allowed for an accumulation of methanogens inside the digester, as the HRT of TPAD1 of 2 days was not enough to avoid washing out the methanogens [22]. However, the presence of fast-growing methanogens (generation time 4–12 h) could not be ruled out either [44], especially when any of the other means for methanogenic inhibition such as lowering the pH and dosing methanogenic inhibitors were not performed [19].

Our experimental results suggested that the TPAD system was beneficial due to an improved hydrolysis and acidogenesis in the first stage, and optimized conditions for methanogenesis in the second stage. Such a system seemed to be sufficiently efficient, mainly at a short total HRT of TPAD up to 14 days, which could reduce the footprint and investment costs. The authors of [45] stated HRT to be a crucial parameter that can influence the efficiency of AD, and an HRT of 30 days allows all types of AD to become more or less the same in terms of biogas production, which makes the more energy-demanding TAD and TPAD less economically interesting. The authors of [42,46] underlined that the first stage of TPAD was the most efficient at 2–3 days, when the total HRT was less than 20 days.

3.2. *Digestate Dewaterability*

3.2.1. Centrifugation

The dewaterability of the digestates from the TAD, MAD and TPAD systems was determined by means of a dewaterability coefficient, which allowed for us to assess the concentration of dry matter in a dewatered digestate sample. Thus, the higher dewaterability coefficient, the better dewatering efficiency (Table 6, Figure 4).

Table 6. Dewaterability coefficient of the digestates from MAD, TAD and TPAD reactors.

Phase	Dewaterability Coefficient, %		
	TAD	MAD	TPAD2
Phase A, at HRT = 19 days	16.1 ± 0.9	13.8 ± 0.7	14.8 ± 0.7
Phase B, at HRT = 13.5 days	17.4 ± 0.9	13.6 ± 0.6	15.7 ± 0.7

Footnote. TAD—thermophilic anaerobic digestion; MAD—mesophilic anaerobic digestion; TPAD—temperature-phased anaerobic digestion; TPAD2—the second stage of TPAD; HRT—hydraulic retention time.

It was found that the difference among dewaterability coefficients was relatively small, but still statistically significant among all types of AD systems at both Phases. Hence, the best dewaterability was determined for the digestate from TAD, followed by TPAD and MAD. Furthermore, decreasing the HRT from 19 to 13.5 days did not decrease the dewaterability; in fact, it was slightly increased for TAD and TPAD.

Specifically, at 19 days of HRT, the digestates' dewaterability was 13.8%, 14.8% and 16.1% for the MAD, TPAD and TAD (Figure 4a), respectively; while, at 13.5 days of HRT, the digestates dewaterability was 13.6%, 15.7% and 17.4% for the MAD, TPAD and TAD (Figure 4b), respectively (Table 6). Therefore, the dewaterability of TAD was 9.8% and 8.1% higher than TPAD at 19 and 13.5 days of HRT, respectively, while the dewaterability of TAD was 21.8% and 14.3% higher than MAD at 19 and 13.5 days of HRT, respectively.

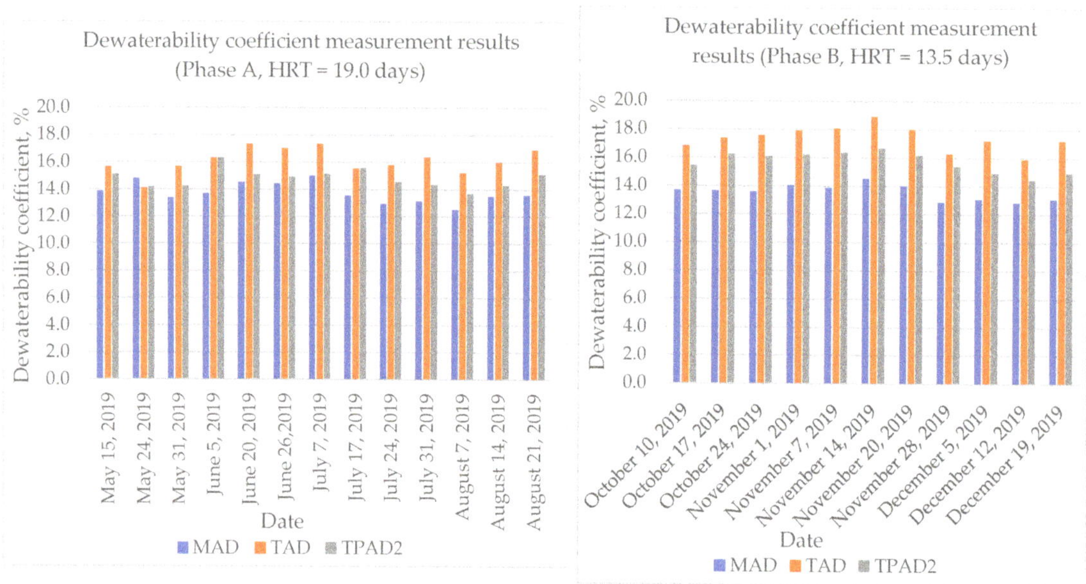

(**a**) (**b**)

Figure 4. Dewaterability coefficient at Phase A (**a**) and at Phase B (**b**).

Hence, despite just a slight effect of the HRT change from 13.5 to 19 days on all types of AD digestate dewaterability, the digestate from TAD showed, continuously, a better performance concerning the ability to "lose" water under the centrifugal forces. The worst quality of digestate after MAD can be explained by a lower degradability of the sludge in terms of VSS (Table 2).

3.2.2. Mechanical Pressing

To the best of our knowledge, the sludge cake concentrations obtained by the mechanical pressing method was in good agreement with the range of results generally achieved in full-scale wastewater treatment plants [47,48]. The ratio between the wet sample and dry cake weight showed how much the digestate could be dewatered. The results of mechanical pressing are depicted in Table 7.

Table 7. The results of mechanical pressing.

Phase	Parameter	TAD	MAD	TPAD2
Phase A, at HRT = 19.0 days	TS of sludge cake, % polymer dose, g/kgTS	25.0 ± 1.0 35.0	26.1 ± 3.8 35.0	25.6 ± 1.7 35.0
Phase B, at HRT = 13.5 days	TS of sludge cake, % polymer dose, g/kgTS	30.8 ± 4.2 30.0	31.4 ± 2.4 30.0	28.7 ± 4.3 32.5

Footnote. TS—total solids; TAD—thermophilic anaerobic digestion; MAD—mesophilic anaerobic digestion; TPAD—temperature-phased anaerobic digestion; TPAD2—the second stage of TPAD; HRT—hydraulic retention time.

In agreement with centrifugation results, the digestate dewaterability did not decrease with HRT. In fact, it was slightly improved after decreasing the HRT from 19 to 13.5 days (Table 7). In addition, the optimal dose of flocculant was slightly lower at the shorter HRT: 35 vs. 30–23.5 g/kgTS for 19 and 13.5 days of HRT, respectively. However, statistically, the obtained results turned out to be insignificant.

The results of different dewaterability measurement methods were quite different, which went along with the literature [31]. However, the trend was similar to another study where TAD-digested sludge showed a better ability to be dewatered and demanded a higher flocculant consumption [49].

The way dewaterability influences the final disposal is straightforward: it is always better when it is as high as possible as by removing the contained water, the sludge reduces in volume, which is beneficial, at least in transportation expenses and any following final disposal starting from old-fashioned landfilling and heading to its reuse in road construction via incineration or direct usage in agriculture [10,29,32].

3.3. Elemental Analysis and Lower Calorific Value

The digested sludge quality was also characterized by the elemental analysis (Table 8).

Table 8. The elemental composition of digested sludge (average values).

Phase	Element, %	TAD	MAD	TPAD1	TPAD2	Substrate
Phase A, at HRT = 19.0 days	N	3.87 ± 0.05	4.23 ± 0.21	4.31 ± 0.05	3.83 ± 0.22	5.37 ± 0.13
	C	25.70 ± 0.27	25.56 ± 0.25	27.32 ± 0.70	24.60 ± 0.19	30.26 ± 0.50
	H	4.39 ± 0.14	4.36 ± 0.12	4.65 ± 0.11	4.29 ± 0.06	4.97 ± 0.19
	S	0.85 ± 0.02	0.84 ± 0.01	0.84 ± 0.01	0.88 ± 0.04	0.77 ± 0.01
	O	65.20 ± 0.48	65.01 ± 0.59	62.88 ± 0.87	66.41 ± 0.51	58.64 ± 0.82
Phase B, at HRT = 13.5 days	N	4.50 ± 0.06	5.18 ± 0.17	5.06 ± 0.07	4.61 ± 0.07	6.50 ± 0.03
	C	30.75 ± 0.53	30.45 ± 0.34	32.58 ± 0.07	29.80 ± 0.59	35.28 ± 0.30
	H	4.77 ± 0.21	4.74 ± 0.18	4.95 ± 0.23	4.63 ± 0.27	5.31 ± 0.20
	S	0.92 ± 0.03	0.98 ± 0.01	0.90 ± 0.03	0.99 ± 0.01	0.81 ± 0.01
	O	59.06 ± 0.79	58.66 ± 0.28	56.52 ± 0.35	59.97 ± 0.80	52.10 ± 0.50

Footnote. TAD—thermophilic anaerobic digestion; MAD—mesophilic anaerobic digestion; TPAD—temperature-phased anaerobic digestion; TPAD1—the first stage of TPAD; TPAD2—the second stage of TPAD; HRT—hydraulic retention time.

Furthermore, the lower calorific value was calculated according to the literature [34] and assessed with respect to the initial value of the substrate LCV (Table 9).

Table 9. Lower calorific value of the digested sludges and substrate.

Phase	TAD		MAD		TPAD1		TPAD2		Substrate	
	LCV, kJ/kg	Loss in LCV, %	LCV, kJ/kg	Loss in LCV, %	LCV, kJ/kg	Loss in LCV, %	LCV, kJ/kg	Loss in LCV, %	LCV, kJ/kg	Loss in LCV, %
Phase A, at HRT = 19.0 days	9157 ± 76	15.4	9111 ± 73	15.8	9742 ± 251	10.0	8750 ± 63	19.2	10,827 ± 155	-
Phase B, at HRT = 13.5 days	11,127 ± 97	13.0	10,949 ± 137	14.3	11,715 ± 29	8.4	10,793 ± 89	15.6	12,783 ± 8.9	-

Footnote. LCV—lower calorific value; TAD—thermophilic anaerobic digestion; MAD—mesophilic anaerobic digestion; TPAD—temperature-phased anaerobic digestion; TPAD1—the first stage of TPAD; TPAD2—the second stage of TPAD; HRT—hydraulic retention time.

During the AD process, part of the substrate organic matter content was biodegraded and converted into methane; thus, reducing the lower energy content of the sludge, here determined by the LCV [50]. The highest LCV decrease was observed in TPAD (around 19% with HRT of 19 days), which supported the highest rate of organics transformation into biogas. In addition, according to the statistical ANOVA test, it was assessed that the obtained LCV data were significantly different only at Phase A (HRT = 19.0 days) and in between TAD–TPAD and MAD–TPAD. This went along with the data on the VS removal rate (Table 2): 32.4% of VS removal at TPAD against 27.2% at TAD and 26.0% at MAD.

The same trend was observed regarding the methane production (Table 4): 233.1 mL/g COD_{added} at TPAD vs. 169.4 mL/g COD_{added} at TAD and 156.0 mL/g COD_{added} at MAD. Which brings us to interesting hypotheses: (1) the longer the HRT, the bigger the difference among the introduced AD systems; (2) the longer the HRT, the bigger the difference between single- and double-stage AD systems.

Considering the sludge cake concentration presented in Table 9, it can be stated that, despite the leftover water content, the real calorific value (related to the wet sludge cake after dewatering) remained quite high, which is important especially when thermal treatment is applied as the final treatment process. As it is known, according to Tanner's triangle, the autothermic process of combustion is highly dependent of the fuel LCV and possible unless the LCV of the digestate is lower than 50% of the loss in the LCV [51].

It was reported that an elemental analysis of the sludge can also be used for the prediction of the dewatered sludge cake TSS concentration [31]. The results depicted in Table 10 had a certain extent of correlation with the solids content of digestate samples after mechanical pressing, shown in Table 7.

Table 10. Sludge cake solids prediction for the digestate after AD and its correlation with mechanical pressing results.

Phase	TAD		MAD		TPAD		Substrate	
	Cake Solids as TSS, %	Correl. Coef.	Cake Solids as TSS, %	Correl. Coef.	Cake Solids as TSS, %	Correl. Coef.	Cake Solids as TSS, %	Correl. Coef.
Phase A, at HRT = 19.0 days	26.5	1.06	23.8	0.91	25.6	1.00	20.4	-
Phase B, at HRT = 13.5 days	23.6	0.77	20.6	0.66	22.9	0.80	16.0	-

Footnote. TSS—total suspended solids; TAD—thermophilic anaerobic digestion; MAD—mesophilic anaerobic digestion; TPAD—temperature-phased anaerobic digestion; correl. coef.—correlation coefficient between the mechanical pressing results (Table 7) and sludge cake solids concentration calculated according to [31]; HRT—hydraulic retention time.

It was noted that at Phase A (HRT = 19.0 days), the correlation coefficient was around 1.0 for all AD systems. At Phase B (HRT = 13.5 days), the correlation coefficient was approximately 20% lower than for the correspondent AD system. It showed that, at a longer HRT, the theoretically calculated prognosis on sludge cake solids concentration was closer (±10%) to the experimental results of the dewatering process by mechanical pressing than at the shorter HRT (lower by 20–30%, on average). This means that at HRTs shorter than 19.0 days, the calculated results on sludge dewaterability properties and based on EA should be verified by laboratory experiments. There might be obtained actual results better than anticipated by theoretical calculations.

3.4. Hygienisation Efficiency Assessment

It is known that sewage sludge contains different types of pathogens, including eggs of parasitic worms, bacteria and viruses. AD is one of the effective methods for the reduction in pathogens to allow the safe application of digested sludge for agriculture [4]. However, depending on the temperature regime, the results of hygienisation may vary: after MAD, the digestate did not meet the requirements that would permit to apply the digestate as a fertilizer to soil; meanwhile, after TAD, the digestate possessed higher pathogenic safety results [52]. Thus, normally, the TAD digestate meets the requirements of Class A biosolids, which are not feasible for MAD [53].

Microbiological analyses were performed to evaluate the potential of digestate to be applied on agricultural fields, directly or after a post-treatment step, which is one of the final disposal applications of digestate [3] (Table 11).

Table 11. Microbiological characterization of the digested sludge concerning the pathogenic safety.

Phase	Microbiological Parameter	WAS from the Feeding Bucket, Stored at +11.5 °C	TAD	MAD	TPAD1	TPAD2
Phase A, at HRT = 19.0 days	Microorganisms cultivated at 22 °C, CFU/g *	2.5×10^6	2.1×10^4	6.2×10^4	3.7×10^4	1.3×10^5
	Microorganisms cultivated at 36 °C, CFU/g *	1.2×10^6	1.4×10^4	7.4×10^4	3.3×10^4	9.1×10^4
	COLI, CFU/g *	8.2×10^4	<1	299	<1	38
	E. coli, (CFU/g *	4.9×10^4	<1	155	<1	<1
	CLO, CFU/g *	2.3×10^4	2.5×10^3	1.3×10^4	1.5×10^4	4.8×10^3
Phase B, at HRT = 13.5 days	Microorganisms cultivated at 22 °C, CFU/g *	2.8×10^7	1.4×10^5	1.2×10^6	4.0×10^5	1.2×10^6
	Microorganisms cultivated at 36 °C, CFU/g *	1.2×10^7	2.2×10^5	9.8×10^5	9.7×10^5	1.7×10^6
	COLI, CFU/g *	3.7×10^4	<1	38	<1	<1
	E. coli, (CFU/g *	2.0×10^3	<1	20	<1	<1
	CLO, CFU/g *	5.0×10^4	1.4×10^3	1.8×10^4	9.2×10^3	4.9×10^3

* Footnote 1. CFU—colony-forming units; TC22 °C—total counts of culturable microorganisms at 22 °C; TC36 °C—total counts of culturable microorganisms at 36 °C; COLI—total counts of coliforms, ECOLI—total counts of *Escherichia coli*, CLO—total counts of *Clostridium perfringens*. Footnote 2. WAS—waste-activated sludge; TAD—thermophilic anaerobic digestion; MAD—mesophilic anaerobic digestion; TPAD—temperature-phased anaerobic digestion; TPAD1—the first stage of TPAD; TPAD2—the second stage of TPAD; HRT—hydraulic retention time.

Table 11 shows that both digestion systems using thermophilic conditions outperformed the mesophilic one. Concerning the mesophilic conditions, the reduction in pathogenic bacteria was less efficient. Decreasing the HRT from 19 to 13.5 days did not impair the pathogenic safety in all evaluated AD systems, since the results could be even better.

The statistics revealed that the only significant difference in microbiological tests was observed for Phase A with 19.0 days of HRT regarding two microbiological parameters of coliforms and *Escherichia coli*, and only in relation to TAD and TPAD towards MAD. The difference between TAD and TPAD was insignificant. TPAD achieved only slightly worse results in comparison with TAD; however, the hygienisation was sufficient for the application of digested sludge to soil, only in the case of Phase A with 19.0 days of HRT. It was also noticed that, though the first stage of TPAD under thermophilic conditions showed a number of coliforms and *Escherichia coli* below the detection level, after changing to mesophilic conditions in the second stage, they appeared again, which might be of concern when defining the HRT of each stage of the double-stage AD system. However, in the TAD digestate, as well as in the TPAD digestate, pathogens were present in significantly lower amounts than after the MAD process. This went along with the results obtained by [49], which stated that after 2 days of the fermenter HRT under thermophilic conditions, some pathogens were not detected, and after 3 days of the fermenter HRT, *Escherichia coli* was completely deactivated. The assured pathogenic safety of TAD-digested sludge and the sludge obtained after the TPAD system with an HRT of the fermenter being long enough for the full deactivation of faecal indicators, allows for the sludge to be directly used in agriculture [5].

3.5. Comparison of Results

All the data obtained were evaluated and placed into Table 12 for a better assessment.

Table 12. Comparison of the obtained data concerning TAD, MAD and TPAD.

Phase	Parameter	TAD	MAD	TPAD1	TPAD2	TPAD
Phase A, at HRT = 19.0 days	VS removal, %	+	-	ND	++	ND
	VFA concentration, mgCOD/L	+	++	-	++	ND
	Methane production, mL/gCOD$_{added}$	++	++	-	++	+++
	Dewaterability coefficient, %	++	-	ND	+	ND
	Polymer dose, g/kgTS	-	-	ND	-	ND
	LCV, kJ/kg	+	+	ND	-	ND
	Cake solids as TSS, %	++	-	ND	+	ND
	Microorganisms cultivated at 22 °C, CFU/g	+++	+	++	-	ND
	Microorganisms cultivated at 36 °C, CFU/g	+++	+	++	-	ND
	COLI, CFU/g	+	-	+	+	ND
	E. coli, CFU/g	+	-	+	+	ND
	CLO, CFU/g	+	-	-	+	ND
	WWTP-LCA *	+	-	ND	ND	++
	SL-LCA **	++	+	ND	ND	0
Phase B, at HRT = 13.5 days	VS removal, %	-	-	ND	+	ND
	VFA concentration, mgCOD/L	+	++	-	++	ND
	Methane production, mL/gCOD$_{added}$	++	++	-	++	+++
	Dewaterability coefficient, %	++	-	ND	+	ND
	Polymer dose, g/kgTS	-	-	ND	+	ND
	LCV, kJ/kg	+	+	ND	-	ND
	Cake solids as TSS, %	++	-	ND	+	ND
	Microorganisms cultivated at 22 °C, CFU/g	++	-	+	-	ND
	Microorganisms cultivated at 36 °C, CFU/g	++	+	+	-	ND
	COLI, CFU/g	-	-	-	-	ND
	E. coli, CFU/g	-	-	-	-	ND
	CLO, CFU/g	+++	-	+	++	-

Footnote 1: "-"—the worst result of all; "+", "++", "+++"—relative estimation in comparison to the worst result (the more "+", the better results compared to "-"-result); ND—no data. Footnote 2: VS—volatile solids; TS—total solids; TSS—total suspended solids; VFA—volatile fatty acids; COD—chemical oxygen demand; LCV—lower calorific value; CFU—colony-forming units; TC22 °C—total counts of culturable microorganisms at 22 °C; TC36 °C—total counts of culturable microorganisms at 36 °C; COLI—total counts of coliforms; ECOLI—total counts of *Escherichia coli*; CLO—total counts of *Clostridium perfringens*; TAD—thermophilic anaerobic digestion; MAD—mesophilic anaerobic digestion; TPAD—temperature-phased anaerobic digestion; TPAD1—the first stage of TPAD; TPAD2—the second stage of TPAD; HRT—hydraulic retention time.* Footnote 3: WWTP-LCA—life cycle assessment of each AD system analysed separately as a part of the whole WWTP with the functional unit of 1 m^3 of treated wastewater (performed only for Phase A; HRT—19.0 days) [2]. ** Footnote 4: SL-LCA—life cycle assessment of each AD system analysed separately as an AD system only with the functional unit of 1 m^3 of produced methane (performed only for Phase A; HRT—19.0 days) [2].

When considering Table 12, all the measured parameters can be split into five groups (for Phases A and B altogether, as there was a negligible difference between the Phases): (1) organic matter degradation efficiency and methane production; (2) process stability (VFA content); (3) sludge quality (dewaterability); (4) final disposal as a fuel (LCV); (5) final disposal as a fertilizer (microbiological parameters). An additional 6th group was assessed for Phase A only—(6) environmental burden (LCA)—, as the LCA was performed only at an HRT of 19 days [2]. The results are depicted in Table 13.

Table 13. Final comparison of TAD, MAD and TPAD.

Phase	Parameter	TAD	MAD	TPAD
Phases A and B	Degradation efficiency and methane production	2	3	1
	Process stability (VFA content)	2	1	1
	Digestate quality (dewaterability)	1	3	2
	Final disposal as a fuel (LCV)	1	1	2
	Final disposal as a fertilizer (pathogen safety)	1	3	2
	5-group average value	1.40	2.20	1.60
Phase A	Environmental burden	1	3	2
	6-group average value	1.33	2.33	1.67

Footnote 1: "1"—one point which relates to the best result; "2"—two points mean the middle-point result; "3"—three points mean the worst result. Footnote 2: VS—volatile solids; TS—total solids; TSS—total suspended solids; VFA—volatile fatty acids; COD—chemical oxygen demand; LCV—lower calorific value; CFU—colony-forming units; TC22 °C—total counts of culturable microorganisms at 22 °C; TC36 °C—total counts of culturable microorganisms at 36 °C; COLI—total counts of coliforms; ECOLI—total counts of *Escherichia coli*; CLO—total counts of *Clostridium perfringens*; TAD—thermophilic anaerobic digestion; MAD—mesophilic anaerobic digestion; TPAD—temperature-phased anaerobic digestion; TPAD1—the first stage of TPAD; TPAD2—the second stage of TPAD; HRT—hydraulic retention time.

Based on Table 13, it can be stated that at both Phases and, correspondently, at both HRTs of 19 and 13.5 days, TAD outperformed. Additionally, the included LCA estimation [2] allowed TAD to obtain more "points" and improve the final mark from 1.40 to 1.33. The difference, according to the five-group-parameter averages, TAD obtained a 0.2-point advantage over TPAD, and TPAD obtained a 0.6-point advantage over MAD, which resulted in a difference between TAD and MAD of up to 0.8. Looking at the six-group average values, it can be claimed that the results were even more improved for TAD and worsened for TPAD and MAD. The TAD advantage over TPAD grew up to 0.34 points, and the TPAD advantage went up to 0.66 points; the overall difference between TAD and MAD went up to 1.0.

It is important to mention that Table 13 represents quite a rough estimation, as only the main characteristics of the AD process were compared. In addition, each characteristic of AD had a different value economically—and ecologically—wise, which has to be considered when making a choice of AD systems for implementation at each WWTP individually. Hence, a bigger number of groups could be presented, and, in its turn, each group (including the introduced ones) could contain more AD parameters. Nevertheless, it gave a good overview of single—and double—stage AD systems according to main process characteristics specifically grouped according to the total AD efficiency, its stability, digestate quality and its possible final disposal.

The obtained data can be compared with the data published earlier—Table 14.

Table 14. Comparison of TAD, MAD and TPAD results with other studies.

Phase	Parameter		TAD	MAD	TPAD	Other Source
Phases A and B	Degradation efficiency as VS decrease, %	Study	22.9–27.2	26.0–32.8	28.8–32.4	-
		Other sources	-	24–34	38–48	[40]
	Specific methane production, mL/gVS$_{dded}$	Study	168–244 *	189–220 *	314–413 *	-
		Other sources	-	111–185	370	[40]
				140	360	[19]
	Process stability as VFA content, mgCOD/L	Study	3.9–4.8	0.3–0.5	0.4–0.5	-
		Other sources	0.87 **	0.16 **	0.31 **	[45]
	Energetic value as LCV loss, kJ/kg	Study	13.0–15.4	14.3–15.8	15.6–19.2	-
		Other sources	16.24 **	16.74 **	16.59 **	[45]

Footnote 1: VS—volatile solids; VFA—volatile fatty acids; COD—chemical oxygen demand; LCV—lower calorific value; TAD—thermophilic anaerobic digestion; MAD—mesophilic anaerobic digestion; TPAD—temperature-phased anaerobic digestion; HRT—hydraulic retention time. * Footnote 2: The average values of specific methane production were recalculated to gVS$_{dded}$ based on data in Table 2. ** Footnote 3: The data presented in the literature source relate to food waste, not sewage sludge.

In addition to Table 14 data, it is needed to mention that [41,54,55] stated that the TPAD process with 15 days of HRT outperforms any of the single-stage systems in terms of dewaterability, though there are still many unsettled issues about the sludge dewaterability measurement and assessment [32]. The same was valid concerning pathogenic safety, with the only exceptional requirement of a minimum HRT of the 1st stage, which should be equal to 3 days. Hence, these studies indicate that TPAD seems to be the most beneficial alternative among other AD systems at a short HRT, similarly as in the presented study.

4. Conclusions

Based on the results obtained in this study, the following conclusions can be drawn:

1. Organic matter removal and methane production experimental data clearly showed that TPAD obtained the best results, followed by TAD and, finally, by MAD.
2. Regarding the dewaterability, the results varied depending on the physical mechanism of the dewatering test. By centrifugation without flocculant addition, the highest dewaterability was obtained by TAD, which was 8.1% and 9.8% higher than TPAD and 14.3% and 21% higher than MAD during both HRTs (13.5 and 19.0 days, respectively). The mechanical pressing results showed the statistical insignificance among the AD systems.
3. The calorific value of the sludge was reduced by 19.2% after TPAD at Phase A with an HRT of 19 days, which was the only statistically significant difference between TPAD and TAD/MAD. At Phase B with an HRT of 13.5, none of the AD systems showed any statistical difference in relation to the other ad systems.
4. The deactivation of pathogens was proven for the TAD digestate regardless of the HRT, but not for the MAD digestate, while TPAD showed different results depending on the HRT. It seems that the HRT of the first stage of TPAD is crucial in relation to the TPAD digestate's pathogenic safety. Hence, the possibility of using the TPAD digestate directly for agricultural purposes might still be a concern.

To sum up the digested quality evaluation, several sludge properties were quantified and compared to aggregate data for making a decision about the suitability of different sludge types for different sludge valorisation routes. It was shown that the TAD digestate can be applied directly in agriculture, while the TPAD digestate might also be used as a fertilizer successfully, depending on the fermenter HRT assuring pathogenic safety. With the highest absolute value of LCV (for dry sludge), MAD was the best for being used as a fuel, preserving a higher portion of organic matter not transformed into biogas, but losing this advantage due to the worst dewaterability in comparison with TAD and TPAD. In terms of the environmental burden, TAD turned out to be the most environmentally friendly one, followed by TPAD and MAD.

In agreement with other studies, it can be stated that the double-stage TPAD system was the most beneficial AD system among the others, allowing a flexible sludge valorisation in different ways. However, its output is highly dependent on: (1) the AD substrate and its characteristics; (2) properly selected operating parameters such as the temperature regime, HRT and OLR.

Author Contributions: Conceptualization, P.J. and I.F.; methodology, P.J. and I.F.; software, J.H. and I.L.; validation, P.J. and I.F.; formal analysis, I.L.; investigation, I.L.; resources, J.H., J.Ř.-A. and I.L.; data curation, I.L.; writing—original draft preparation, I.L.; writing—review and editing, J.Ř.-A., P.J. and I.F.; supervision, P.J. and I.F.; project administration, P.J.; funding acquisition, P.J. and I.F. All authors have read and agreed to the published version of the manuscript.

Funding: This research was funded by the European Union's Horizon 2020 research and innovation programme under the Marie Skłodowska-Curie grant agreement no 676070. This communication reflects only the authors' view and the Research Executive Agency of the EU is not responsible for any use that may be made of the information it contains.

Institutional Review Board Statement: Not applicable.

Informed Consent Statement: Not applicable.

Data Availability Statement: The data presented in the study are available upon request from the correspondent author.

Acknowledgments: I.F. is grateful to the Government of Catalonia (Consolidated Research Group 2017 SGR 1029).

Conflicts of Interest: The authors declare no conflict of interest. The funders had no role in the design of the study; in the collection, analyses, or interpretation of data; in the writing of the manuscript, or in the decision to publish the results.

Abbreviations

AD	anaerobic digestion
COD	chemical oxygen demand
EA	elemental analysis
GC	gas chromatography
HRT	hydraulic retention time
LCA	life cycle assessment
LCV	lower calorific value
MAD	mesophilic anaerobic digestion
OLR	organic loading rate
PE	people equivalent
SL-LCA	life cycle assessment of each AD system analysed separately as an AD system only with the functional unit of 1 m^3 of produced methane
sCOD	soluble chemical oxygen demand
tCOD	total chemical oxygen demand
TAD	thermophilic anaerobic digestion
TPAD	temperature-phased anaerobic digestion
TPAD1	the first stage (fermenter) of TPAD
TPAD2	the second stage of TPAD
TS	total solids
TSS	total suspended solids
V	reactor working volume
VFA	volatile fatty acid
VS	volatile solids
VSS	volatile suspended solids
WAS	waste-activated sludge
WRRF	water resource recovery facility
WWTP	wastewater treatment plant
WWTP-LCA	life cycle assessment of each AD system analysed separately as a part of the whole WWTP with the functional unit of 1 m^3 of treated wastewater

References

1. Zhao, G.; Garrido-Baserba, M.; Reifsnyder, S.; Xu, J.-C.; Rosso, D. Comparative energy and carbon footprint analysis of biosolids management strategies in water resource recovery facilities. *Sci. Total Environ.* **2019**, *665*, 762–773. [CrossRef]
2. Lanko, I.; Flores, L.; Garfí, M.; Todt, V.; Posada, J.A.; Jenicek, P.; Ferrer, I. Life Cycle Assessment of the mesophilic, thermophilic, and temperature-phased anaerobic digestion of sewage sludge. *Water* **2020**, *12*, 3140. [CrossRef]
3. Kominko, H.; Gorazda, K.; Wzorek, Z. Potentiality of sewage sludge-based organo-mineral fertilizer production in Poland considering nutrient value, heavy metal content and phytotoxicity for rapeseed crops. *J. Environ. Manag.* **2019**, *248*, 109283. [CrossRef] [PubMed]
4. Pauline, A.L.; Joseph, K. Hydrothermal carbonization of oily sludge for solid fuel recovery – investigation of chemical characteristics and combustion behaviour. *J. Anal. Appl. Pyrolysis* **2021**, *157*, 105235. [CrossRef]
5. Seleiman, M.F.; Santanen, A.; Mäkelä, P.S.A. Recycling sludge on cropland as fertilizer – Advantages and risks. *Resour. Conserv. Recycl.* **2020**, *155*, 104647. [CrossRef]
6. Chen, T.; Qiu, X.; Feng, H.; Yin, J.; Shen, D. Solid digestate disposal strategies to reduce the environmental impact and energy consumption of food waste-based biogas systems. *Bioresour. Technol.* **2021**, *325*, 124706. [CrossRef]
7. Neczaj, E.; Grosser, A. 31—Biogas production by thermal hydrolysis and thermophilic anaerobic digestion of waste-activated sludge. *Ind. Munic. Sludge* **2019**, 741–781. [CrossRef]

8. Ferrer, I.; Vazquez, F.; Font, X. Long term operation of a thermophilic anaerobic reactor: Process stability and efficiency at decreasing sludge retention time. *Bioresour. Technol.* **2010**, *101*, 2972–2980. [CrossRef] [PubMed]
9. Markis, F.; Baudez, J.-C.; Parthasarathy, R.; Slatter, P.; Eshtiaghi, N. The apparent viscosity and yield stress of mixtures of primary and secondary sludge: Impact of volume fraction of secondary sludge and total solids concentration. *Chem. Eng. J.* **2016**, *288*, 577–587. [CrossRef]
10. Liu, H.; Li, X.; Zhang, Z.; Nghiem, L.D.; Gao, L.; Wang, Q. Semi-continuous anaerobic digestion of secondary sludge with free ammonia pretreatment: Focusing on volatile solids destruction, dewaterability, pathogen removal and ist implications. *Water Res.* **2021**, *202*, 117481. [CrossRef]
11. Jiang, X.; Lyu, Q.; Bi, L.; Xie, Y.; Ji, G.; Huan, C.; Xu, L.; Yan, Z. Improvement of sewage sludge anaerobic digestion through synergistic effect combined trace elements enhancer with enzyme pretreatment and microbial community response. *Chemosphere* **2022**, *286*, 131356. [CrossRef] [PubMed]
12. Micolucci, F.; Gottardo, M.; Pavan, P.; Cavinato, C.; Bolzonella, D. Pilot scale comparison of single and double-stage thermophilic anaerobic digestion of food waste. *J. Clean. Prod.* **2018**, *171*, 1376–1385. [CrossRef]
13. Chen, Y.; Jiang, S.; Yuan, H.; Zhou, Q.; Gu, G. Hydrolysis and acidification of waste activated sludge at different pHs. *Water Res.* **2007**, *41*, 683–689. [CrossRef] [PubMed]
14. Wan, J.; Jing, Y.; Zhang, S.; Angelidaki, I.; Luo, G. Mesophilic and thermophilic alkaline fermentation of waste activated sludge for hydrogen production: Focusing on homoacetogenesis. *Water Res.* **2016**, *102*, 524–532. [CrossRef] [PubMed]
15. Wu, S.-L.; Wei, W.; Ni, B.-J. Enhanced methane production from anaerobic digestion of waste activated sludge through preliminary pretreatment using calcium hypochlorite. *J. Environ. Manag.* **2021**, *295*, 113346. [CrossRef] [PubMed]
16. Liang, Z.; Shen, N.; Lu, C.; Chen, Y.; Guan, Y. Effective methane production from waste activated sludge in anaerobic digestion via formic acid pretreatment. *Biomass Bioenergy* **2021**, *151*, 106176. [CrossRef]
17. Qasim, S.R. *Wastewater Treatment Plants: Planning, Design, and Operation*, 2nd ed.; Routledge: Boca Raton, FL, USA, 1999. [CrossRef]
18. Lin, R.; Cheng, J.; Ding, L.; Murphy, J.D. Improved efficiency of anaerobic digestion through direct interspecies electron transfer at mesophilic and thermophilic temperature ranges. *Chem. Eng. J.* **2018**, *350*, 681–691. [CrossRef]
19. Ruffino, B.; Campo, G.; Cerutti, A.; Scibilia, G.; Lorenzi, E.; Zanetti, M. Comparative analysis between a conventional and a temperature-phased anaerobic digestion system: Monitoring of the process, resources transformation and energy balance. *Energy Convers. Manag.* **2020**, *223*, 113463. [CrossRef]
20. Li, X.; Guo, S.; Peng, Y.; He, Y.; Wang, S.; Li, L.; Yhao, M. Anaerobic digestion using ultrasound as pretreatment approach: Changes in waste activated sludge, anaerobic digestion performances and digestive microbial populations. *Biochem. Eng. J.* **2018**, *139*, 139–145. [CrossRef]
21. Oles, J.; Dichtl, N.; Niehoff, H. Full scale experience of two stage thermophilic/mesophilic sludge digestion. *Water Sci. Technol.* **1997**, *36*, 449–456. [CrossRef]
22. Qin, Y.; Higashimori, A.; Wu, L.-J.; Hojo, T.; Kubota, K.; Li, Y.-Y. Phase separation and microbial distribution in the hyperthermophilic-meso-type temperature-phased anaerobic digestion (TPAD) of waste activated sludge (WAS). *Bioresour. Technol.* **2017**, *245*, 401–410. [CrossRef]
23. Ma, D.; Li, A.; Zhang, L.; Wang, D.; Ji, G. Mechanical compression assisted conductive drying of thin-film dewatered sewage sludge: Process performance, heat and mass transfer behavior. *Waste Manag.* **2021**, *126*, 41–51. [CrossRef]
24. Sakarika, M.; Stavropoulos, K.; Kopsahelis, A.; Koutra, E.; Zafiri, C.; Kornaros, M. Two-stage anaerobic digestion harnesses more energy from the co-digestion of end-of-life dairy products with agro-industrial waste compared to the single-stage process. *Biochem. Eng. J.* **2020**, *153*, 107404. [CrossRef]
25. Nguyen, P.-D.; Truong Tran, N.-S.; Nguyen, T.-T.; Dang, B.-T.; Thi Le, M.-T.; Bui, X.-T.; Mukai, F.; Kobayashi, H.; Ngo, H.-H. Long-term operation of the pilot scale two-stage anaerobic digestion of municipal biowaste in Ho Chi Minh City. *Sci. Total Environ.* **2021**, *766*, 142562. [CrossRef]
26. Shi, Z.; Zhao, R.; Wan, J.; Li, B.; Shen, Y.; Zhang, S.; Luo, G. Metagenomic analysis reveals the fate of antibiotic resistance genes in two-stage and one-stage anaerobic digestion of waste activated sludge. *J. Hazard. Mater.* **2021**, *406*, 124595. [CrossRef]
27. Yuan, D.; Wang, Y.; Qian, X. Variations of internal structure and moisture distribution in activated sludge with stratified extracellular polymeric substances extraction. *Water ResearcInternational Biodeterior. Biodegrad.* **2017**, *116*, 1–9. [CrossRef]
28. Zhang, W.; Xu, Y.; Dong, B.; Dai, X. Characterizing the sludge moisture distribution during anaerobic digestion process through various approaches. *Sci. Total Environ.* **2019**, *675*, 184–191. [CrossRef] [PubMed]
29. Wei, H.; Gao, B.; Ren, J.; Li, A.; Yang, H. Coagulation/flocculation in dewatering of sludge: A review. *Water Res.* **2018**, *143*, 608–631. [CrossRef]
30. Abu-Orf, M.M.; Dentel, S.K. Effect of mixing on the rheological characteristics of conditioned sludge: Full-scale studies. *Water Sci. Technol.* **1997**, *36*, 51–60. [CrossRef]
31. Svennevik, O.K.; Beck, G.; Rus, E.; Westereng, B.; Higgins, M.; Solheim, O.E.; Nilsen, P.J.; Horn, S.J. CNash-A novel parameter predicting cake solids of dewatered digestates. *Water Res.* **2019**, *158*, 350–358. [CrossRef] [PubMed]
32. Wu, B.; Dai, X.; Chai, X. Critical review on dewatering of sewage sludge: Influential mechanism, conditioning technologies and implications to sludge re-utilizations. *Water Res.* **2020**, *180*, 115912. [CrossRef]
33. Astals, S.; Venegas, C.; Peces, M.; Jofre, J.; Lucena, F.; Mata-Alvarez, J. Balancing hygienization and anaerobic digestion of raw sewage sludge. *Water Res.* **2012**, *46*, 6218–6227. [CrossRef] [PubMed]

34. Nzihou, J.F.; Hamidou, S.; Bouda, M.; Koulidiati, J.; Segda, B.G. Using Dulong and Vandralek formulas to estimate the calorific heating value of a household waste model. *Int. J. Sci. Eng. Res.* **2014**, *5*, 1878–1883.
35. Technical Committee CEN/TC 230 "Water Analysis". *ISO 6222: 1999 Water Quality—Enumeration of Culturable Micro-Organisms—Colony Count by Inoculation in a Nutrient Agar Culture Medium*; ISO (The International Organization for Standardization): Geneva, Switzerland, 1999.
36. Technická Normalizační Komise. *ČSN 75 7835 (757835): 2009 Jakost Vod—Stanovení Termotolerantních Koliformních Bakterií a Escherichia coli [Water Quality—Enumeration of Thermotolerant Coliform Bacteria and Escherichia coli]*; Úřad pro Technickou Normalizaci, Metrologii a Státní Zkušebnictví: Prague, Czech Republic, 2009.
37. Technická Normalizační Komise. *ČSN 75 7837 (757837): 2010 Jakost Vod—Stanovení Koliformních Bakterií v Nedesinfikovaných Vodách [Water Quality—Enumeration of Coliform Bacteria in Non-Disinficated Waters]*; Úřad pro Technickou Normalizaci, Metrologii a Státní Zkušebnictví: Prague, Czech Republic, 2010.
38. EU, C.D. 98/83/EC of Council of 3rd of November 1998 on the quality of water intended for human consumption. *Off. J. Eur. Commun.* **1998**, *L330*, 32–54.
39. APHA. *Standard Methods for the Examination of Water and Wastewater*, 22rd ed.; Rice, E.W., Baird, R.B., Eaton, A.D., Clesceri, L.S., Eds.; American Public Health Association (APHA): Washington, DC, USA; American Water Works Association (AWWA): Denver, FL, USA; Water Environment Federation (WEF): Chicago, IL, USA, 2012.
40. Ge, H.; Jensen, P.D.; Batstone, D.J. Increased temperature in the thermophilic stage in temperature phased anaerobic digestion (TPAD) improves degradability of waste activated sludge. *J. Hazard. Mater.* **2020**, *187*, 355–361. [CrossRef] [PubMed]
41. Fernández-Rodríguez, J.; Pérez, M.; Romero, L.I. Semicontinuous Temperature-Phased Anaerobic Digestion (TPAD) of Organic Fraction of Municipal Solid Waste (OFMSW). Comparison with single-stage processes. *Chem. Eng. J.* **2016**, *285*, 409–416. [CrossRef]
42. Amodeo, C.; Hattou, S.; Buffier, P.; Benbelkacem, H. Temperature phased anaerobic digestion (TPAD) of organic fraction of municipal solid waste (OFMSW) and digested sludge (DS): Effect of different hydrolysis conditions. *Waste Manag.* **2021**, *126*, 21–29. [CrossRef] [PubMed]
43. Cao, Z.; Hülsemann, B.; Wüst, D.; Illi, L.; Oechsner, H.; Kruse, A. Valorization of maize silage digestate from two-stage anaerobic digestion by hydrothermal carbonization. *Energy Convers. Manag.* **2020**, *222*, 113218. [CrossRef]
44. Weimer, P.J. Manipulating ruminal fermentation: A microbial ecological perspective. *J. Anim. Sci.* **1998**, *76*, 3114–3122. [CrossRef]
45. Xiao, B.; Qin, Y.; Zhang, W.; Wu, J.; Qiang, H.; Liu, J.; Li, Y.-Y. Temperature-phased anaerobic digestion of food waste: A comparison with single-stage digestions based on performance and energy balance. *Bioresour. Technol.* **2018**, *249*, 826–834. [CrossRef]
46. Wu, L.-L.; Kobayashi, T.; Li, Y.-Y.; Xu, K.-Q. Comparison of single-stage and temperature-phased two-stage anaerobic digestion of oily food waste. *Energy Convers. Manag.* **2015**, *106*, 1174–1182. [CrossRef]
47. Toutian, V.; Barjenbruch, M.; Loderer, C.; Remy, C. Pilot study of thermal alkaline pretreatment of waste activated sludge: Seasonal effects on anaerobic digestion and impact on dewaterability and refractory COD. *Water Res.* **2020**, *182*, 115910. [CrossRef]
48. Cai, Y.; Zheng, Z.; Wang, X. Obstacles faced by methanogenic archaea originating from substrate-driven toxicants in anaerobic digestion. *J. Hazard. Mater.* **2021**, *403*, 123938. [CrossRef]
49. Lloret, E.; Pastor, L.; Pradas, P.; Pascual, J.A. Semi full-scale thermophilic anaerobic digestion (TAnD) for advanced treatment of sewage sludge: Stabilization process and pathogen reduction. *Chem. Eng. J.* **2013**, *232*, 42–50. [CrossRef]
50. Menon, U.; Suresh, N.; George, G.; Ealias, A.M.; Saravanakumar, M.P. A study on combined effect of Fenton and Free Nitrous Acid treatment on sludge dewaterability with ultrasonic assistance: Preliminary investigation on improved calorific value. *Chem. Eng. J.* **2020**, *382*, 123035. [CrossRef]
51. Flaga, A. The aspects of sludge thermal utilization. In *Conference Paper*; Institute of Heat Engineering and Air Protection, Cracow University of Technology: Cracow, Poland, 2010.
52. Hupfauf, S.; Winkler, A.; Wagner, A.O.; Podmirseg, S.M.; Insam, H. Biomethanation at 45 °C offers high process efficiency and supports hygienisation. *Bioresour. Technol.* **2020**, *300*, 122671. [CrossRef]
53. Fu, B.; Wang, Y.; Liu, H.; Chen, Y.; Jiang, Q.; Liu, H. Balancing energy production and pathogen removal during temperature phased anaerobic digestion of sewage sludge. *Fresenius Environ. Bull.* **2014**, *23*, 1643–1649.
54. Riau, V.; Angeles De La Rubia, M.; Perez, M. Temperature-phased anaerobic digestion (TPAD) to obtain class A biosolids: A semi-continuous study. *Bioresour. Technol.* **2010**, *101*, 2706–2712. [CrossRef] [PubMed]
55. Nabaterega, R.; Kumar, V.; Khoei, S.; Eskicioglu, C. A review on two-stage anaerobic digestion options for optimizing municipal wastewater sludge treatment process. *J. Environ. Chem. Eng.* **2021**, *9*, 105502. [CrossRef]

water

MDPI

Article

Toward the Adoption of Anaerobic Digestion Technology through Low-Cost Biodigesters: A Case Study of Non-Centrifugal Cane Sugar Producers in Colombia

Oscar Mendieta [1,2,*], Liliana Castro [3,*], Erik Vera [4], Jader Rodríguez [1] and Humberto Escalante [2]

1 Corporación Colombiana de Investigación Agropecuaria-AGROSAVIA, Centro de Investigación Tibaitatá, 14 km Vía Mosquera Bogotá, Mosquera 250040, Colombia; jrodriguezc@agrosavia.co
2 Grupo de Investigación en Tecnologías de Valorización de Residuos y Fuentes Agrícolas e Industriales para la Sustentabilidad Energética-INTERFASE, Escuela de Ingeniería Química, Universidad Industrial de Santander—UIS, Carrera 27, Calle 9 Ciudad Universitaria, Bucaramanga 680001, Colombia; escala@uis.edu.co
3 Centro de Estudios e Investigaciones Ambientales-CEIAM, Escuela de Ingeniería Química, Universidad Industrial de Santander—UIS, Carrera 27, Calle 9 Ciudad Universitaria, Bucaramanga 680001, Colombia
4 Grupo de Investigación en Ciencia de Materiales Biológicos y Semiconductores-CIMBIOS, Escuela de Física, Universidad Industrial de Santander—UIS, Carrera 27, Calle 9 Ciudad Universitaria, Bucaramanga 680001, Colombia; erik.vera@correo.uis.edu.co
* Correspondence: omendieta@agrosavia.co (O.M.); licasmol@saber.uis.edu.co (L.C.)

Abstract: Anaerobic digestion using low-cost biodigesters (LCB) is a promising alternative for Colombian producers of non-centrifugal cane sugar (NCS). Since the integration of anaerobic digestion technology in this agro-industry is novel, it is critical to understand the factors that affect the acceptance behavior of such technology by NCS producers to develop future policies that promote the adoption of sustainable energy alternatives. This study aimed to analyze NCS producers' behavioral intention to use LCB by utilizing an extended technology acceptance model (TAM). Data from a survey of 182 producers were used to evaluate the proposed model empirically. The extended TAM accounted for 78% of the variance in producers' behavioral intention to use LCB. Thus, LCB acceptability could be fairly precisely predicted on the basis of producers' intentions. This study's findings contribute to research on the TAM and provide a better understanding of the factors influencing NCS producers' behavioral intention to use LCB. Furthermore, this approach can assist policymakers at the local and global levels, given that NCS is produced in various developing countries worldwide.

Keywords: anaerobic digestion acceptance; structural equation model; energy policy; sustainable energy technology; rural development

Citation: Mendieta, O.; Castro, L.; Vera, E.; Rodríguez, J.; Escalante, H. Toward the Adoption of Anaerobic Digestion Technology through Low-Cost Biodigesters: A Case Study of Non-Centrifugal Cane Sugar Producers in Colombia. *Water* **2021**, *13*, 2566. https://doi.org/10.3390/w13182566

Academic Editor: Bing-Jie Ni

Received: 24 August 2021
Accepted: 15 September 2021
Published: 17 September 2021

1. Introduction

The traditional production of non-centrifugal cane sugar (NCS) from sugarcane is found in many developing countries [1]. For example, NCS production in Colombia is the country's second-largest agricultural sector, after coffee, with 220,000 ha of sugarcane cultivation. Over 350,000 families participate in this agro-industry, which generates 287,000 direct jobs and employs approximately 12% of the country's economically active rural population [2]. However, the NCS agro-industry has historically faced many challenges related to low agricultural and processing productivity, substandard product quality, and producer organization issues, all of which have hampered entry into new markets. The latter is reflected in the fact that a sizable proportion of producers and workers live in poverty. The sugarcane used in NCS production generates approximately 24.6% of its mass in organic waste, resulting in negative environmental impacts. This agro-industry generates 3.9 million tons of waste per year (according to Colombian production conditions). Crop residues are normally burned in the open air, or wastewater is dumped into bodies of water, resulting in odors, greenhouse gas (GHG) emissions, and water and soil pollution [3]. The issue of finding alternatives to the waste generated by the agro-industry is currently being addressed.

Previous research has revealed that anaerobic digestion (AD) technology contributes to sustainability and a circular bioeconomy in different agro-industrial sectors [4,5]. Among alternative conversion methods, AD was shown to be the most sustainable biomass-to-energy technology for municipal waste management, with 34 indicators utilized within the context of three-dimensional sustainability (economic, environmental, and social) [6]. The increase in the number of agricultural biogas plants is a manifestation of the ongoing energy transition and an opportunity to achieve the objectives related to the implementation of the circular bioeconomy [7]. Specifically, recent research has shown that the main residues in NCS production (i.e., agricultural crop residues and sugarcane scum), when managed by AD, achieve synergy for bioenergy production [8]. A maximum methane yield of $0.276\ Nm^3\ CH_4 \cdot kg^{-1}\ VS_{added}$ was obtained for co-digestion, which constitutes an efficient form to take advantage of biomass. Subsequently, the feasibility of the AD process in low-cost biodigesters with a tubular configuration was determined, achieving a specific biogas production of $0.132\ m^3 \cdot kg^{-1}$ VS with a methane content of 50.4%, and profitability indicators confirmed the economic viability of the technology for NCS producers [9]. Then, through a life-cycle analysis, the environmental sustainability of the AD technology integrated into the NCS production process was documented, highlighting the mitigation of the eutrophication impact categories up to 99% [10]. As a result, renewable fuel (biogas) and a biofertilizer (digestate) were obtained to assist the NCS sector in transitioning to sustainability and a circular bioeconomy.

Although the technical feasibility and the economic [9] and environmental [10] benefits of integrating low-cost biodigesters (LCB) into the NCS sector have been established, the producers' acceptance of this technology has not been assessed. The overwhelming majority of AD studies have been approached from a technical, environmental, and economic perspective. Nevertheless, social acceptance, which is a pillar of sustainability and circular bioeconomy for technology adoption, has not been considered. The social acceptance of sustainable energy production systems should be addressed before the implementation stage, which would help stakeholders create policies that help spread the technology. Understanding the fundamentals of the factors influencing NCS producers' acceptance of LCB will assist in making successful decisions to disseminate AD technology. Technology acceptance is considered a very important issue in the field of agriculture. Many theories and models have been used to accurately and systematically identify factors affecting innovation acceptance [11].

The technology acceptance model (TAM) and theory of planned behavior under perceived production risks were combined to analyze the factors affecting farmers' intentions to adopt information and communication technologies in intensive shrimp production in Vietnam [12]. The TAM has also been used to promote smart farming by Iran farmers [13], for which constructs were established that directly impact behavioral intention. On the other hand, a systematic review was carried out to adopt digital agricultural technologies to transform current agricultural systems toward sustainability, based on the diffusion of innovation theory (DIT) [14]. The constructs employed in the TAM, integrated with the perceived innovation characteristics of the DIT, provided an even more robust model than either of the two models alone [15,16]. TAM and DIT have been used in a variety of disciplines such as water resource management [17,18], sociology [19,20], and agriculture [21]. However, the TAM and DIT could be combined with additional variables to improve the model's predictive capacity. This could offer an approach toward the acceptance of anaerobic digestion technology/innovation. Despite the vast amount of research undertaken using the TAM in additional scientific fields, the literature has not explored its application in the waste management sector or, in particular, in AD technology. This study attempts to fill this gap by understanding the factors that affect the acceptance of AD technology by NCS producers in Colombia toward the adoption of LCB through an extended TAM. Overall, this study contributes to research on the TAM and, for the first time, to an understanding of NCS producers' LCB acceptance behavior. The TAM integrated with DIT, as well as two additional constructs (perceived self-efficacy and facilitating conditions) used in this work, is a more comprehensive model for analyzing LCB acceptance but has not been used in previous waste management studies. This empirical research supports the validity of this integrated model and constitutes an important contribution to adopting new sustainable energy technologies for developing countries. The findings of this study can help policymakers encourage NCS producers to use LCB and overcome some of the bottlenecks that arise during the implementation of such technology as a waste management tool and its contribution to the circular bioeconomy.

2. Materials and Methods

2.1. Extended Technology Acceptance Model

The technology acceptance model (TAM) is a popular model widely used in numerous studies on the acceptance and usage of information technologies [22]. The TAM determines technology acceptance (actual use) based on behavioral intention to use. Behavioral intention to use is influenced by attitude toward use and the direct and indirect effects of perceived usefulness and perceived ease of use. The degree to which an individual believes that using a particular system will improve their job performance is described as perceived usefulness. In contrast, perceived ease of use refers to how much a person thinks it would be free of effort to use the system. Both constructs, perceived usefulness and perceived ease of use, jointly affect the attitude toward use, while perceived ease of use directly impacts perceived usefulness.

The TAM model proposed by Venkatesh and Davis [23] eliminates the component attitude toward use (which previously mediated some of the impact of perceived usefulness and perceived ease of use). As with the original TAM, the actual use of technology is determined by behavioral intention to use. The latter is the focus of this study to predict AD technology acceptance. According to the modified Venkatesh and Davis [23] model, the following hypotheses were proposed:

Hypotheses 1 (H1). *Perceived usefulness has a direct effect on behavioral intention to use.*

Hypotheses 2 (H2). *Perceived ease of use has a direct effect on behavioral intention to use.*

Hypotheses 3 (H3). *Perceived ease of use has a direct effect on perceived usefulness.*

Rogers [24] proposed the DIT, which is now widely used to define and justify adopting technologies, ranging from agricultural tools to organizational developments. Five innovation characteristics are included in the DIT: relative advantage, compatibility, complexity, trialability, and observability. However, previous research has found that only relative advantage, compatibility, and complexity are consistently linked to innovation acceptance [25]. Perceived usefulness is similar to relative advantage, while perceived ease of use is similar to complexity. On the other hand, compatibility refers to how well an innovation fits into potential adopters' existing values, past experiences, and needs. Potential adopters will shy away from a new idea if it does not match society's traditions and values. The DIT is similar to the TAM in that it emphasizes the psychological and social influences on an individual's behavioral intention to adopt new technology [22]. According to this modified model, and taking compatibility as a construct in the TAM, the following hypotheses were proposed:

Hypotheses 4 (H4). *Compatibility has a direct effect on perceived usefulness.*

Hypotheses 5 (H5). *Compatibility has a direct effect on behavioral intention to use.*

Previous research has demonstrated that the TAM's predictive ability could be enhanced by incorporating additional variables [26]. More precisely, it has been shown that the concepts of perceived self-efficacy and facilitating conditions are significant determinants of behavioral intention to use new technologies. Perceived self-efficacy is a construct that explains how an individual assesses their own ability to perform a task successfully [27]. Perceived self-efficacy is a key predictor of behavior because completing a job depends not only on a person's knowledge but also on the person's belief in their ability to complete it. Concerning LCBs' acceptance, perceived self-efficacy would indicate how NCS producers would perceive their ability, experience, skills, and expertise required for the use of AD technology integrated into the NCS process. As a result, perceived self-efficacy plays an important role in LCB acceptance, leading to the following hypothesis:

Hypotheses 6 (H6). *Perceived self-efficacy has a direct effect on behavioral intention to use.*

Facilitating conditions refer to an individual's belief that an adequate organizational and technical infrastructure level exists to support the system's use [28]. In the context of LCB, facilitating conditions include various approaches to meeting producers' needs, such as related training programs and workshops, technical consultants, and technical guidelines. Therefore, the following hypothesis was included in the model:

Hypotheses 7 (H7). *Facilitating conditions have a direct effect on behavioral intention to use.*

In this study, the TAM was integrated with the DIT and two additional constructs (perceived self-efficacy and facilitating conditions) to model acceptance of LCB by NCS producers. TAM provided the constructs of perceived usefulness, perceived ease of use, and behavioral intention to use. The DIT was used to elicit the concept of compatibility. The model also incorporates the other two external constructs, namely, perceived self-efficacy and facilitating conditions. The model used in this study is depicted in Figure 1; the dotted line in the figure denotes the study's scope, which is to predict Colombian NCS producers' behavioral intention to accept LCBs.

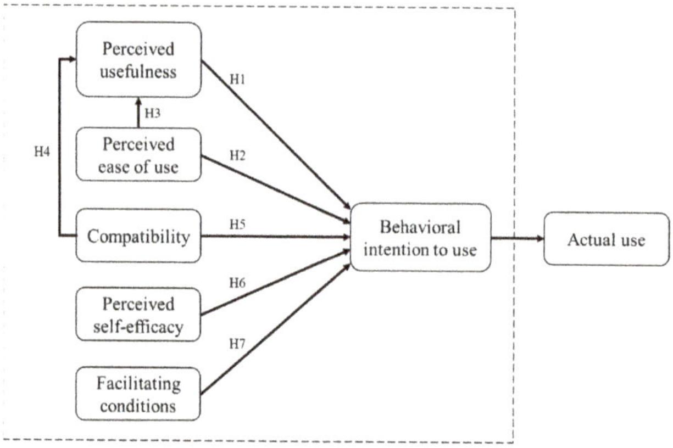

Figure 1. Extended technology acceptance model for LCB acceptance by NCS producers.

2.2. *Survey*

A structured survey was used to elicit data on the sample's sociodemographic characteristics, and a series of items was used to assess the constructs. The measures for all constructs were adapted from previously validated instruments and contextualized for LCB acceptance. Perceived usefulness, perceived ease of use, and behavioral intention to use were derived from prior research on the TAM [22,29,30]. The compatibility measures were adapted from Rogers [24], while the LCB perceived self-efficacy items were tailored from Venkatesh and Davis [29]. Venkatesh et al. [28] provided the facilitating conditions items modified for use in the current study.

The study's data collection method was through interviews with NCS producers. The survey consisted of two sections. The first section consisted of six questions capturing NCS producers' and production units' characteristics, including sex, age, formation, NCS production experience, sugarcane area, and yearly NCS production. The second section collected data on producers' perceptions of the model's variables. This section included 16 items that assessed the six constructs (perceived usefulness, perceived ease of use, compatibility, perceived self-efficacy, facilitating conditions, and behavioral intention to use). All variables were measured using a five-point Likert scale (1: totally disagree, 2: disagree, 3: I have no idea, 4: agree, and 5: totally agree). The survey was initially reviewed by three NCS producers and three local extension agents to adjust the elements and constructs used in the research and explain the instrument's terminology, content, and general design. The survey application was then tested with 12 NCS producers to ensure they understood the questions, technical terms, and measurement scales. The researchers' observations and feedback from producers and local extension officers resulted in minor revisions to the survey instructions, the rewording of several items, and an explanation for several technical terms. Once the instrument was reviewed and adjusted, the definitive information was collected for the investigation.

2.3. Study Area and Sample

Figure 2 shows the area planted with sugar cane for the production of NCS in Colombia distributed in the country by departments. The following departments are highlighted: Boyacá, Santander, Nariño, Antioquia, Cundinamarca, Tolima, Huila, and Cauca. The annual average air temperature for cultivation is 26–32 °C during the day and 13–17 °C at night [31]. The AD technology-based alternative for waste management is still in its early stages of adoption, which is why this research is so important for the government and private entities seeking its early implementation.

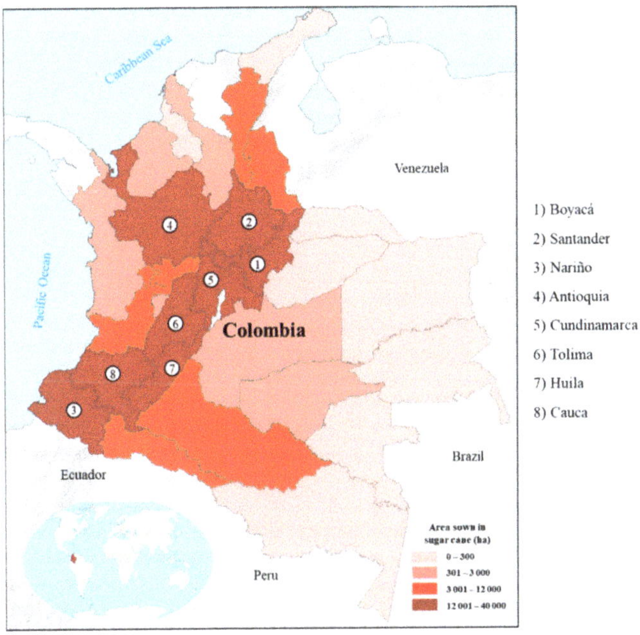

Figure 2. Map of the study area: sugarcane plantations in Colombia used to produce NCS; adapted from AGRONET [32].

Snowball and convenience sampling techniques were used to collect data for this study. Initially, respondents were chosen on the basis of proximity and ease of access to the researchers during the survey, using convenience sampling as a nonprobability sampling technique with an accidental sampling technique. The first wave of respondents recommended potential NCS producers to include in this study via a chain-referral system [33].

2.4. Data Analysis

Linear structural relations (LISREL 10.20 [34]), a conventional and multilevel structural equation modeling program, was used for descriptive statistics and data modeling. The measurement instrument's reliability and validity were assessed using reliability, discriminant, and convergent validity criteria. Cronbach's alpha coefficient was used to determine the survey instrument's reliability [35]. Convergent and discriminant constructs validity were examined using average variance extracted analysis [36] and Pearson's correlation coefficient evaluated using Evans [37] guidelines (correlation levels: negligible = 0.00–0.19, weak = 0.20–0.39, moderate = 0.40–0.59, strong = 0.60–0.79, very strong = 0.80–1.00). The kurtosis and skewness of each construct were computed to verify the distribution of the data (normality). Additionally, exploratory factor analysis was used to determine the convergent validity of each construct.

The hypothesized relationships were tested using structural equation modeling. Path analysis was used to test the hypothesized relationships between variables and the theoretical model presented in Figure 1 based on multiple regression analyses. According to the model developed in the theoretical framework, two regression models were used to investigate the relationships between the variables. The relationship between perceived usefulness, the dependent variable, and compatibility and perceived ease of use, the independent variables, was examined in model 1. Model 2 examined the relationship between behavioral intention to use and perceived usefulness, perceived ease of use, compatibility, perceived self-efficacy, and facilitating conditions.

Evaluation of model fit is not as straightforward in structural equation modeling as in statistical approaches based on error-free variables. Because no single statistical significance test can reliably identify the correct model given the sample data, it is necessary to consider multiple criteria and evaluate model fit using multiple concurrent measures. For each estimation procedure, many goodness-of-fit indices are given to determine if the model is consistent with the empirical evidence. The present study used the significance test (χ^2 test statistic) and descriptive goodness-of-fit measures to evaluate the model's fit. For the latter, the following indices were used [38]: root-mean-square error of approximation (RMSEA), standardized root-mean-square residual (SRMR), non-normed fit index (NNFI), comparative fit index (CFI), goodness-of-fit index (GFI), and adjusted goodness-of-fit index (AGFI).

3. Results and Discussion

3.1. Descriptive Statistics

In total, 187 responses were collected from Colombia's major NCS-producing departments. Five of the cases that responded provided insufficient information and were discarded, leaving 182 questionnaires completed. MacCallum [39] suggested 100–200 cases to obtain factor solutions that are adequately stable and closely correspond to the population factors.

Table 1 shows the demographic profile of the respondents. Agro-industrial activities of the processing of sugarcane to produce NCS have been led by the male gender. Women in rural areas of developing countries are frequently prevented from working outside the home and on family farms due to cultural, social, and religious norms [40]. The majority of respondents were in the age category 46–55 years. Meanwhile, almost 30% of the respondents corresponded to a population entering older adulthood (>56 years). This aging phenomenon is currently facing the NCS agro-industry, in which a slow generational change is perceived. Most NCS producers had a good education level; only 6.59% had no formal education at all, 39.01% had an elementary school education (5 years of schooling), 26.37% had completed high school (11 years of education), and some NCS producers (28.02%) had obtained a college degree (education spanning more than 11 years).

Table 1. Demographic attributes of the respondents (N = 182).

Variables	Frequency	Percentage
Gender		
Female	13	7.14
Male	169	92.86
Age (mean = 49.46 years)		
Less than 35 years	30	16.48
From 35 to 45 years	28	15.38
From 46 to 55 years	73	40.11
From 56 to 65 years	33	18.13
More than 65 years	18	9.89
Education		
No education at all	12	6.59
Elementary school	71	39.01
High school graduate	48	26.37
Some college	51	28.02
NCS production experience (mean = 29.14 years)		
Less than 10 years	18	9.89
From 10 to 20 years	41	22.53
From 21 to 30 years	15	8.24
More than 30 years	108	59.34
Area sowed in sugar cane (mean = 21.68 ha)		
Less than 5 ha	13	7.14
From 5 to 25 ha	136	74.73
From 26 to 50 ha	25	13.74
From 51 to 75 ha	5	2.75
More than 75 ha	4	2.20
Annual NCS production (mean = 150.72 t)		
Less than 25 t	12	6.59
From 25 to 50 t	35	19.23
From 51 to 100 t	59	32.42
From 101 to 200 t	45	24.73
More than 200 t	31	17.03
NCS producer location		
Boyacá	25	13.74
Santander	27	14.84
Nariño	18	9.89
Antioquia	22	12.09
Cundinamarca	34	18.68
Tolima	20	10.99
Huila	22	12.09
Cauca	14	7.69

The majority of NCS producers surveyed had extensive experience producing NCS (average 29.14 years); a sizable percentage (59.34%) had more than 30 years of experience. Because NCS production has been a family tradition, the link with the agro-industry is established at an early age. The vast majority (74.73%) of NCS producers owned between 5 and 25 ha of sugarcane. As a result, the majority of producers in the study area were small-scale. NCS production averaged 150.72 t per year, with most producers producing between 51 and 100 t (sugarcane yield ranges between 4 and 11 $t \cdot year^{-1} \cdot ha^{-1}$). The location of the NCS producers revealed a dispersion across the Colombian territory, encompassing the primary NCS producing departments for this study.

3.2. *Reliability and Validity Testing*

The initial statistical results of the data collected with the measurement instrument are shown in Table 2.

Table 2. Statistical analysis of the data obtained with the measuring instrument.

Construct	Compatibility	Perceived Ease of Use	Perceived Usefulness	Perceived Self-Efficacy	Facilitating Conditions	Behavioral Intention to Use
			Covariance matrix			
Compatibility	0.84					
Perceived ease of use	0.28	0.62				
Perceived usefulness	0.41	0.27	0.74			
Perceived self-efficacy	0.25	0.23	0.26	0.79		
Facilitating conditions	0.32	0.22	0.33	0.33	0.88	
Behavioral intention to use	0.51	0.38	0.47	0.42	0.49	0.90
			Pearson's correlation matrix			
Compatibility	1.00					
Perceived ease of use	0.46 (0.000)	1.00				
Perceived usefulness	0.53 (0.000)	0.50 (0.000)	1.00			
Perceived self-efficacy	0.31 (0.004)	0.41 (0.000)	0.36 (0.000)	1.00		
Facilitating conditions	0.36 (0.000)	0.36 (0.000)	0.42 (0.000)	0.40 (0.000)	1.00	
Behavioral intention to use	0.62 (0.000)	0.64 (0.000)	0.63 (0.000)	0.55 (0.000)	0.58 (0.000)	1.00
			Statistics			
Mean	4.19	4.57	4.41	4.12	4.08	4.30
Standard deviation	0.92	0.65	0.82	0.97	0.97	0.90
Skewness	1.02	1.36	1.30	1.09	1.26	1.22
Kurtosis	0.45	1.22	1.05	1.13	1.38	0.92
Cronbach's alpha	0.85	0.93	0.83	0.83	0.89	-

The average variance extracted (AVE) square root is presented on the covariance matrix's main diagonal. The values in parentheses in the Pearson correlation matrix show the significance of the values (two-tailed).

Cronbach's alpha coefficient of the measurement instrument had an average value of 0.87, indicating a high internal consistency level, since it was higher than the recommended minimum of 0.8 for basic research purposes [34]. Therefore, the reliability of the survey instrument was confirmed. Furthermore, the average variance extracted (AVE) square root was much larger than all other cross-correlations for the sample. The square root of the AVE for all measures exceeded the recommended level of 0.5 [35] (ranged from 0.62 to 0.90), indicating that the hypothesized constructs accounted for more than half of the variability observed in the items. Pearson's correlation matrix results showed that the independent variables were positively correlated from moderate to strong with behavioral intention to use (ranged from 0.55 to 0.64). In contrast, perceived usefulness was linked from weak to moderate with compatibility and perceived ease of use. However, all variables were significantly correlated with the behavioral intention to use and each other at $p < 0.01$ (values in parentheses). The mean values of all variables were above four and with standard deviations between 0.65 and 0.97, indicating that the vast majority of respondents agreed with or tended to agree with the variables' statements. Furthermore, the distribution of the data was mainly of moderate normality since the absolute values of skewness and kurtosis averaged 1.18, which is in the range of 1 to 2.3 reported by Lei and Lomax [41], except for compatibility kurtosis, which was considered a normal distribution (<1). As a result, the data analysis using structural equation modeling was adequate.

Constructs with the measurable indicators and the factor loading are shown in Table 3. The factor loading coefficient is the correlation coefficient between the constructs and the factor analysis measures. This analysis revealed that all measures had factor loadings greater than 0.6 (if a correlation with an instrument measuring the same construct is >0.50, convergent validity is generally considered adequate [42]), thus verifying the convergent validity of each construct. Thus, the criteria above confirmed the measurement instrument's reliability, convergent validity, and discriminant validity.

Table 3. Factor analysis between the constructs and the measurable variables.

Construct	Measures	Factor Loading
Compatibility	Using the low-cost biodigesters is compatible with most aspects of an NCS mill	0.83
	Using low-cost biodigesters to produce bioenergy and biofertilizer is compatible with the environment and climate of this region	0.89
	Using the low-cost biodigesters for the benefit of NCS production is consistent with the financial situation of the process	0.71
Perceived ease of use	Learning to operate the low-cost biodigesters would be easy for me	0.94
	The interaction with the low-cost biodigesters would be easy for me to understand	0.86
	I would find the low-cost biodigesters easy to use	0.91
Perceived usefulness	Using the low-cost biodigesters would save time and money	0.72
	The low-cost biodigesters would support critical aspects in an NCS mill	0.82
	I would find the low-cost biodigesters useful in an NCS mill	0.81
Perceived self-efficacy	I could use the low-cost biodigesters if there were no one around to tell me what to do as I go	0.64
	I could use the low-cost biodigesters if I saw someone else using them before trying them myself	0.86
	I could use the low-cost biodigesters if someone showed me how to do it first	0.90
Facilitating conditions	I have the resources necessary to use the low-cost biodigesters	0.88
	I have enough knowledge to use the low-cost biodigesters	0.84
	Given the resources, opportunities, and knowledge it takes to use the low-cost biodigesters, it would be easy for me to use it	0.83
Behavioral intention to use	Assuming I had access to the low-cost biodigesters, I would intend to use it	-

The experience of the producers in NCS production confirmed the maturity of the studied sector (Table 1). Likewise, the low level of non-schooling, associated with the acceptance of the variables investigated, allowed us to discern a concern among NCS producers based not only on survival but also on environmental conservation. Therefore, producers perceived that AD technology could be a solution for sustainable waste management and agro-industry benefit.

3.3. Model Fit Evaluation

The maximum likelihood was the fit function used for the structural equation models. It is consistent and efficient, does not depend on the scale, and is normally distributed if the observed variables are moderately normal [43]. The extent to which the specified models fit the empirical data is shown in Table 4. The degrees of freedom (df) were 24 and 90 for models 1 and 2, respectively. Therefore, the χ^2 test statistic associated with the significance test (p-value) demonstrated the good fit of the models to the data.

The root-mean-square error of approximation (RMSEA) is a statistic that indicates the population's approximate fit and, hence, the difference caused by approximation [44]. RMSEA values were zero in both models; thus, it was inferred that they fit the population approximately well. Furthermore, the lower boundary (left side) of the 90% confidence interval for RMSEA was zero for both models. The standardized root-mean-square residual index (SRMR) is an overall badness-of-fit measure based on the fitted residuals first divided by the standard deviations [45]. Models were found to have an SRMR within the good model's fit range. The other descriptive measures used to evaluate the fit of the models (NNFI, CFI, GFI, and AGFI) presented good fit values according to the literature reviewed.

Considering the previous results on the statistical indices, the adjustment of the proposed models was validated.

Table 4. Results of the model fit evaluation.

Fit Measure	Good Fit	Reference	Model 1	Model 2
χ^2 p-value	$0 \leq \chi^2 \leq 2df$ $0.05 < p \leq 1.00$	[46]	17.31 0.84	79.37 0.78
RMSEA	$0 \leq RMSEA \leq 0.05$	[44]	0.00	0.00
SRMR	$0 \leq SRMR \leq 0.05$	[45]	0.02	0.03
NNFI	$0.97 \leq NNFI \leq 1.00$	[47]	1.00	1.00
CFI	$0.97 \leq CFI \leq 1.00$	[48]	1.00	1.00
GFI	$0.95 \leq GFI \leq 1.00$	[49]	0.98	0.95
AGFI	$0.90 \leq GFI \leq 1.00$ close to GFI	[50]	0.96	0.92

χ^2: chi-square. *df*: degrees of freedom (model 1 = 24, model 2 = 90). RMSEA: root-mean-square error of approximation. SRMR: standardized root-mean-square residual. NNFI: non-normed fit index. CFI: comparative fit index. GFI: goodness-of-fit index. AGFI: adjusted goodness-of-fit index.

3.4. Hypothesis Testing

The results of structural equation modeling are shown in Table 5. The unstandardized coefficients of the independent variables showed positive correlations with their dependent variables. The standard error averaged 0.073 and 0.111 for the independent variables in models 1 and 2, respectively, which indicates an appropriate estimate. In models 1 and 2, the standard error was lower (on average, 0.0215). The standard error shows how precisely the parameter's value was estimated; a smaller standard error denotes a more accurate estimate. Furthermore, models 1 and 2 had variance errors of 0.0826 and 0.180, respectively, sufficient for establishing the parameters. From the Z-values, it is possible to reject the null hypothesis and accept the hypotheses proposed for each model (Z-values greater than 1.96 at a significance level of 5%).

Table 5. Multiple regression analysis results.

Independent Variable	Unstandardized Coefficients		Standardized Coefficients	Z-Values	p-Values
	β	Standard Error	β		
Model 1:	Dependent variable: Perceived usefulness				
Model statistics	Errorvar = 0.0826, R^2 = 0.741, Standard error = 0.0214, Z-value = 3.84, p-value = 0.00				
Compatibility	0.48	0.072	0.64	6.66	0.000
Perceived ease of use	0.27	0.073	0.29	3.68	0.000
Model 2:	Dependent variable: Behavioral intention to use				
Model statistics	Errorvar = 0.180, R^2 = 0.777, Standard error = 0.0216, Z-value = 8.32, p-value = 0.00				
Compatibility	0.25	0.112	0.21	2.23	0.026
Perceived ease of use	0.26	0.093	0.18	2.82	0.005
Perceived usefulness	0.41	0.189	0.26	2.19	0.029
Perceived self-efficacy	0.28	0.0877	0.19	3.14	0.002
Facilitating conditions	0.25	0.0721	0.21	3.41	0.001

The combined hypotheses' path analysis results are shown in Figure 3. Compatibility and perceived ease of use accounted for 74% of the variance in perceived usefulness in model 1. Additionally, compatibility had the greatest effect on perceived usefulness (H4) and was strongly supported (β = 0.64, Z = 6.66, $p < 0.001$). On the other hand, H3 achieved a lower β compared to H4. However, its contribution was also significantly supported (β = 0.29, Z = 3.68, $p < 0.001$). Compatibility, perceived ease of use, perceived usefulness, perceived self-efficacy, and facilitating conditions explained 78% of the variance of behavioral intention to use (model 2).

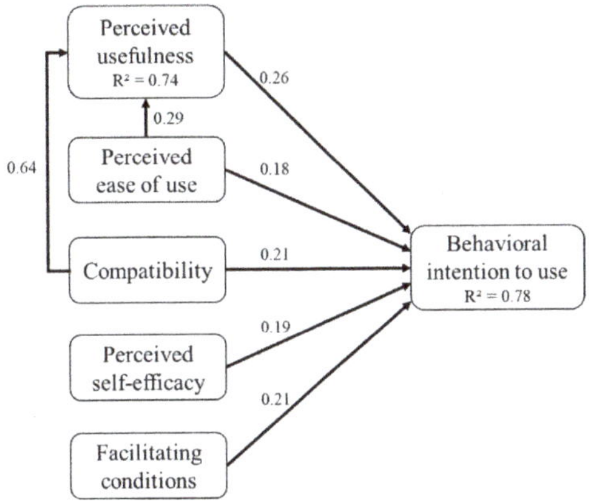

Figure 3. Illustration of the study's empirical findings following the proposed model.

In model 2, the proposed Hypotheses H2 and H5 were corroborated, indicating that perceived ease of use and compatibility also directly affect behavioral intention to use. In this model, compatibility obtained a β greater than that of perceived ease of use (0.21 > 0.18), but both effects (H2 and H5) were supported ($Z > 1.96$, $p < 0.05$). Similarly, the effect of perceived usefulness on behavioral intention to use (H1) was verified ($\beta = 0.26$, $Z = 2.19$, $p < 0.05$), obtaining the highest β coefficient among the proposed constructs on behavioral intention to use. H6 examined the relationship between perceived self-efficacy and behavioral intention to use. As expected, there was a significant correlation ($\beta = 0.19$, $Z = 3.14$, $p < 0.01$), confirming this hypothesis. H7 was associated with the effect of facilitating conditions on behavioral intention to use. H7's path was significant ($\beta = 0.21$, $Z = 3.41$, $p < 0.01$). Therefore, H7 was supported. Overall, it was determined that all path coefficients were statistically significant.

There were indirect effects of some explanatory variables on the dependent variables, in addition to the direct effects summarized in Figure 3. These effects resulted from multiplying all of the direct effects from the explanatory variable through the causal path (e.g., compatibility on perceived usefulness) to the final dependent variable (e.g., behavioral intention to use). Table 6 summarizes the total effect of each explanatory variable on the dependent variables, including direct and indirect effects. The strongest overall effect was for compatibility on the behavioral intention to use LCB (0.38), followed by perceived ease of use and perceived usefulness (both 0.26), facilitating conditions (0.21), and perceived self-efficacy (0.19).

Table 6. Variables' direct, indirect, and total effects on dependent variables.

Variable	Effect on					
	Perceived Usefulness			Behavioral Intention to Use		
	Direct	Indirect	Total	Direct	Indirect	Total
Compatibility	0.64	-	0.64	0.21	0.17	0.38
Perceived ease of use	0.29	-	0.29	0.18	0.08	0.26
Perceived usefulness				0.26	-	0.26
Perceived self-efficacy				0.19	-	0.19
Facilitating conditions				0.21	-	0.21

The total effect is equal to the sum of the direct and indirect effects. The direct effects occur without the intervention of any other variable in the model (i.e., all significant beta values from Figure 3). The indirect effects were calculated by multiplying the coefficients of each independent variable by the coefficients of the related dependent variables.

The extended TAM developed in this study was successfully applied in the context of LCB acceptance by NCS producers, demonstrating the model's robustness. The results indicate that an extended TAM model can capture some of the unique contextual characteristics of NCS producers in terms of LCB acceptance. The findings suggest that producers' intentions accurately predict the factors that affect the acceptance of AD technology by NCS producers, confirmed by pathway analysis (Figure 3), showing that all seven initial hypotheses were supported (Tables 5 and 6). Thus, behavioral intention is significantly affected by compatibility, perceived ease of use, perceived usefulness, perceived self-efficacy, and facilitating conditions.

Similar findings have been reported in previous studies examining the role of farmers' intentions in accepting new technologies through TAM. For instance, when analyzing the intention of sugar beet farmers to accept drip irrigation using the TAM with eight constructs [21], it was possible to account for 59.3% of the variation in farmers' behavioral intentions. In another study, when explaining the acceptance and use of biological control for rice stem borer [16], the TAM explained 78% of the variance in the behavioral intention; however, the perceived ease of use construct was not significant. In contrast, this construct was substantial in the present study, and the two-part model explained 74% of the perceived usefulness variance and accounted for 78% of behavioral intention to use.

The TAM is a well-known technology acceptance theory, and evidence suggests it holds in the agricultural context. However, a new model must be developed for each sector to explain why a specific population accepts a certain technology; hence, the results of this work confirm its importance.

The most significant effect on behavioral intention to use was compatibility (H5). This result corroborates other hypotheses stating the critical role of compatibility with other innovations to explain why the new technology is adopted [22,24]. Compatibility has been crucial in selecting an appropriate rice stem borer management [51] and adopting precision agriculture technology [15]. Compatibility may have had a high priority in the current study because LCB is an on-site waste management technology for bioenergy and biofertilizer production tailored to the needs of the NCS producer. Producers' understanding that LCB is consistent with the farm's environmental conditions, process needs, and affordability of the technology allows them to have a high perception of the advantages of LCB. As a result, they will be more receptive to using LCB on their farms. Compatibility appears to be the most important determinant of LCB success, and, as such, it must be considered when promoting and implementing a technology transfer program. On the other hand, several research studies have found that compatibility directly impacts perceived usefulness [16,52]. By knowing the advantages of LCB in advance, producers are more likely to consider the usefulness of LCB if they consider it compatible with most of their farm conditions and NCS processing activities (supporting H4).

In previous studies [23], perceived usefulness was consistently a strong predictor of usage intention, and similar findings were found in this study (H1). Overall, respondents found LCB useful for waste management, even though the technology has yet to be implemented. Currently, LCB is a useful solution for organic waste generated in various neighboring agro-industries that have benefited economically and environmentally from the technology [53,54]. Perhaps due to these personal experiences, NCS producers are more receptive to technology acceptance, as the utility benefits of LCB outweigh the disadvantages of traditional waste management methods such as landfills or land disposal.

Like the perceived usefulness, the perceived ease of use, as one of the main constructs of the original TAM, showed a strong determinant on the behavioral intention of use (H2). Perceived ease of use refers to the degree to which an NCS producer "feels and perceives" the use of an LCB effortlessly. Previous research has indicated that improving perceived ease of use increases producers' readiness to accept a new technology [21]. Additionally, in the pre-implementation test, perceived ease of use directly and significantly affected behavioral intention to use (little or no direct experience with a particular system) [26]. In the present study, the producers perceived the use of LCB as easy. Therefore, perceived ease of use is a critical variable to consider during the early stages of dissemination, such as the current study, because the success of technological diffusion is highly dependent on it. Consistent with previous research findings [55,56], it was found that perceived ease of use had a substantial positive effect on the perceived utility of LCB implementation (supporting H3). This result indicates that if NCS producers can easily use the LCB, they will consider it more useful.

Another novel result is the effect of facilitating conditions on producers' behavioral intention to use LCB in their practices (H7). Facilitating conditions represent a relatively recent concept used in technology acceptance studies, but they are critical for potential research applications [16]. The findings for this factor support the view that NCS producers who frequently receive extension services and training would welcome any opportunity to promote the use of LCB on their farms and support efforts to disseminate this technology.

The concept of perceived self-efficacy is central to social learning theory. As an individual factor, perceived self-efficacy reflects an individual's beliefs about their ability to complete specific tasks [27]. In the context of the NCS agro-industry, individuals' perceptions of their knowledge, skill, and capability for LCB application are reflected in their perceived self-efficacy. According to a systematic literature review of TAM studies, perceived self-efficacy is the first most commonly used external construct [57]. In this respect, the role of perceived self-efficacy is key to understanding the behavioral intention of NCS producers to use LCB (supporting H6). This finding could be explained by the fact that people with high self-efficacy believe they can perform well using technology [58]. As a result, they are more likely to try the technology and continue to evaluate its benefits.

4. Conclusions

Anaerobic digestion is considered the future of renewable energies with a sustainable approach. In this sense, social acceptance must be analyzed to disseminate the LCB without barriers in agricultural sectors as NCS producers. In this study, the extended technology acceptance model (TAM) predicted the behavioral intention of non-centrifugal cane sugar (NCS) producers to use low-cost digesters (LCB) in Colombia. The TAM was successfully extended (in terms of model fit) by including three external factors relevant to analyze the acceptance of LCB: perceived self-efficacy, facilitating conditions, and compatibility as antecedents of behavioral intention to use. Path analysis showed that the seven hypotheses proposed for the extended TAM were adequately supported with Z-values greater than 1.96 at a significance level of 5%. The model's constructs explained 74% of the variance in perceived usefulness and 78% in behavioral intention to use LCB. The unexplained variance of the dependent variable implies that other variables can also be found within the framework to increase the explanation level. For example, subjective norms and personal relationships that seem to be about behavioral intention toward the acceptance of LCB can be suitable determinants. In this work, the behavioral intention was explained with five factors (i.e., compatibility, perceived self-efficacy, facilitating conditions, perceived usefulness, and perceived ease of use). However, additional factors, such as demographic characteristics of producers (e.g., gender, age, and education) and farm structure characteristics (land size, sugarcane yield, etc.), also exist. These additional factors can also influence the behavioral intention to use the AD technology. Considering this perspective, future research needs to include such elements to build a comprehensive model while maintaining conciseness. Lastly, Colombian NCS producers tended to concur with accepting the integration of anaerobic digestion (AD) technology with the NCS agro-industry. There is a sense of urgency surrounding implementing these AD systems in the NCS agro-industry, which would avoid significant environmental impacts while also generating economic benefits and social inclusion. The findings of this study contribute to a new understanding of NCS producers' perspectives on AD and serve as a guide for developing strategies and resource management for developing countries' technology diffusion policies.

Author Contributions: Conceptualization, O.M. and E.V.; methodology, O.M.; software, E.V.; validation, E.V.; formal analysis, O.M.; investigation, O.M.; resources, J.R.; data curation, E.V.; writing—original draft preparation, L.C.; writing—review and editing, O.M.; visualization, O.M.; supervision, H.E.; project administration, J.R.; funding acquisition, L.C. All authors have read and agreed to the published version of the manuscript.

Funding: This research received no external funding.

Institutional Review Board Statement: Not applicable.

Informed Consent Statement: Not applicable.

Acknowledgments: The authors would like to acknowledge the support of the Ministerio de Ciencia Tecnología e Innovación (Minciencias) convocatoria 757 de 2016, Corporación Colombiana de Investigación Agropecuaria (AGROSAVIA), Federación Nacional de Productores de Panela (FEDEPANELA), Instituto Técnico Agrícola de Cáchira (Norte de Santander-Colombia), and Universidad Industrial de Santander (UIS) for the development of this work.

Conflicts of Interest: The authors declare no conflict of interest. The funders had no role in the design of the study; in the collection, analyses, or interpretation of data; in the writing of the manuscript, or in the decision to publish the results.

References

1. Jaffé, W. Non centrifugal cane sugar (NCS) (panela, jaggery, gur, muscovado) process technology and the need of its innovation. *Panela Monit.* **2014**, *6*, 8–17.
2. MADR. NCS Agroindustrial Chain, Directorate of Agricultural and Forestry Chains, Ministry of Agriculture and Rural Development 2019. Available online: https://sioc.minagricultura.gov.co/Panela/Documentos/2019-12-30%20Cifras%20Sectoriales.pdf (accessed on 20 August 2021).
3. Guerrero, M.; Escobar, J. The technical efficiency of non-centrifugal cane sugar production. *J. Technol.* **2015**, *14*, 107–116. Available online: https://revistas.unbosque.edu.co/index.php/RevTec/article/view/1853 (accessed on 20 August 2021).
4. Mancini, E.; Raggi, A. A review of circularity and sustainability in anaerobic digestion processes. *J. Environ. Manag.* **2021**, *291*, 112695. [CrossRef]
5. Ghaleb, A.A.S.; Kutty, S.R.M.; Salih, G.H.A.; Jagaba, A.H.; Noor, A.; Kumar, V.; Almahbashi, N.M.Y.; Saeed, A.A.H.; Saleh Al-dhawi, B.N. Sugarcane Bagasse as a Co-Substrate with Oil-Refinery Biological Sludge for Biogas Production Using Batch Mesophilic Anaerobic Co-Digestion Technology: Effect of Carbon/Nitrogen Ratio. *Water* **2021**, *13*, 590. [CrossRef]
6. Khan, I.; Kabir, Z. Waste-to-energy generation technologies and the developing economies: A multi-criteria analysis for sustainability assessment. *Renew. Energy* **2020**, *150*, 320–333. [CrossRef]
7. Jain, A.; Sarsaiya, S.; Awasthi, M.K.; Singh, R.; Rajput, R.; Mishra, U.C.; Chen, J.; Shi, J. Bioenergy and bio-products from bio-waste and its associated modern circular economy: Current research trends, challenges, and future outlooks. *Fuel* **2022**, *307*, 121859. [CrossRef]
8. Mendieta, O.; Castro, L.; Rodríguez, J.; Escalante, H. Synergistic effect of sugarcane scum as an accelerant co-substrate on anaerobic co-digestion with agricultural crop residues from non-centrifugal cane sugar agribusiness sector. *Bioresour. Technol.* **2020**, *303*, 122957. [CrossRef]
9. Mendieta, O.; Castro, L.; Rodríguez, J.; Escalante, H. Management and valorization of waste from a non-centrifugal cane sugar mill via anaerobic co-digestion: Technical and economic potential. *Bioresour. Technol.* **2020**, *316*, 123962. [CrossRef]
10. Mendieta, O.; Castro, L.; Escalante, H.; Garfí, M. Low-cost anaerobic digester to promote the circular bioeconomy in the non-centrifugal cane sugar sector: A life cycle assessment. *Bioresour. Technol.* **2021**, *326*, 124783. [CrossRef]
11. Taherdoost, H. A review of technology acceptance and adoption models and theories. *Procedia Manuf.* **2018**, *22*, 960–967. [CrossRef]
12. Ulhaq, I.; Pham, N.T.A.; Le, V.; Pham, H.C.; Le, T.C. Factors influencing intention to adopt ICT among intensive shrimp farmers. *Aquaculture* **2021**, *547*, 737407. [CrossRef]
13. Ronaghi, M.H.; Forouharfar, A. A contextualized study of the usage of the Internet of things (IoTs) in smart farming in a typical Middle Eastern country within the context of Unified Theory of Acceptance and Use of Technology model (UTAUT). *Technol. Society.* **2020**, *63*, 101415. [CrossRef]
14. Shang, L.; Heckelei, T.; Gerullis, M.K.; Börner, J.; Rasch, S. Adoption and diffusion of digital farming technologies-integrating farm-level evidence and system interaction. *Agric. Syst.* **2021**, *190*, 103074. [CrossRef]
15. Aubert, B.A.; Schroeder, A.; Grimaudo, J. IT as enabler of sustainable farming: An empirical analysis of farmers' adoption decision of precision agriculture technology. *Decis. Support Syst.* **2012**, *54*, 510–520. [CrossRef]
16. Sharifzadeh, M.S.; Damalas, C.A.; Abdollahzadeh, G.; Ahmadi-Gorgi, H. Predicting adoption of biological control among Iranian rice farmers: An application of the extended technology acceptance model (TAM2). *Crop Prot.* **2017**, *96*, 88–96. [CrossRef]

17. Ignacio, J.J.; Malenab, R.A.; Pausta, C.M.; Beltran, A.; Belo, L.; Tanhueco, R.M.; Promentilla, M.A.; Orbecido, A. A perception study of an integrated water system project in a water scarce community in The Philippines. *Water* **2019**, *11*, 1593. [CrossRef]
18. Chfadi, T.; Gheblawi, M.; Thaha, R. Public Acceptance of Wastewater Reuse: New Evidence from Factor and Regression Analyses. *Water* **2021**, *13*, 1391. [CrossRef]
19. Rajaee, M.; Hoseini, S.M.; Malekmohammadi, I. Proposing a socio-psychological model for adopting green building technologies: A case study from Iran. *Sustain. Cities. Soc.* **2019**, *45*, 657–668. [CrossRef]
20. Chen, C.F.; Xu, X.; Arpan, L. Between the technology acceptance model and sustainable energy technology acceptance model: Investigating smart meter acceptance in the United States. *Energy Res. Soc. Sci.* **2017**, *25*, 93–104. [CrossRef]
21. Valizadeh, N.; Rezaei-Moghaddam, K.; Hayati, D. Analyzing Iranian Farmers' Behavioral Intention towards Acceptance of Drip Irrigation Using Extended Technology Acceptance Model. *J. Agric. Sci. Technol.* **2020**, *22*, 1177–1190.
22. Davis, F.D.; Bagozzi, R.P.; Warshaw, P.R. User acceptance of computer technology: A comparison of two theoretical models. *Manag. Sci.* **1989**, *35*, 982–1003. [CrossRef]
23. Venkatesh, V.; Davis, F. A theoretical extension of the technology acceptance model: Four longitudinal field studies. *Manag. Sci.* **2000**, *46*, 186–204. [CrossRef]
24. Rogers, E.M. *Diffusion of Innovation*, 5th ed.; Free Press: New York, NY, USA, 2003.
25. Agarwal, R.; Prasad, J. A conceptual and operational definition of personal innovativeness in the domain of information technology. *Inf. Syst. Res.* **1998**, *9*, 204–215. [CrossRef]
26. Wu, J.H.; Wang, S.C. What drives mobile commerce: An empirical evaluation of the revised technology acceptance model. *Inf. Manag.* **2005**, *42*, 719–729. [CrossRef]
27. Marakas, G.; Johnson, R.; Clay, P. The evolving nature of the computer self-efficacy construct: An empirical investigation of measurement construction, validity, reliability and stability over time. *J. Assoc. Inf. Syst.* **2007**, *8*, 16–46. [CrossRef]
28. Venkatesh, V.; Morris, M.; Davis, G.; Davis, F. User acceptance of information technology: Toward a unified view. *MIS Quart.* **2003**, *27*, 425–478. [CrossRef]
29. Venkatesh, V.; Davis, F. A model of the antecedents of perceived ease of use: Development and test. *Decis. Sci.* **1996**, *27*, 451–481. [CrossRef]
30. Venkatesh, V. Determinants of perceived ease of use: Integrating control, intrinsic motivation, and emotion into the technology acceptance model. *Inf. Syst. Res.* **2000**, *11*, 342–365. [CrossRef]
31. López, J. *Agronomic Management of the Sugarcane Crop for NCS in Antioquia*; Colombian Corporation for Agricultural Research: Bogotá, Colombia, 2015.
32. AGRONET. Sugarcane for NCS Production. Ministry of Agriculture and Rural Development 2014. Available online: https://www.agronet.gov.co/Documents/Ca%C3%B1a%20Panelera.pdf (accessed on 20 August 2021).
33. Palinkas, L.A.; Horwitz, S.M.; Green, C.A.; Wisdom, J.P.; Duan, N.; Hoagwood, K. Purposeful sampling for qualitative data collection and analysis in mixed method implementation research. *Adm. Policy Ment. Health* **2015**, *42*, 533–544. [CrossRef] [PubMed]
34. Jöreskog, K.; Sörbom, D. LISREL 10.20 Student Edition (July 2019), Scientific Software International. Available online: https://ssicentral.com/index.php/products/lisrel/ (accessed on 20 August 2021).
35. Henson, R. Understanding internal consistency reliability estimates: A conceptual primer on coefficient alpha. *Meas. Eval. Couns. Dev.* **2001**, *34*, 177–189. [CrossRef]
36. Zaiţ, A.; Bertea, P. Methods for testing discriminant validity. *J. Mark. Manag.* **2011**, *9*, 217–224.
37. Evans, J. Linear correlation. In *Straightforward Statistics for the Behavioral Sciences*; Brooks/Cole Publishing Company: Pacific Grove, CA, USA, 1996; pp. 127–158.
38. Schermelleh-Engel, K.; Moosbrugger, H.; Müller, H. Evaluating the fit of structural equation models: Tests of significance and descriptive goodness-of-fit measures. *MPR-Online* **2003**, *8*, 23–74.
39. MacCallum, R.C.; Widaman, K.F.; Zhang, S.; Hong, S. Sample Size in Factor Analysis. *Psychol. Methods* **1999**, *4*, 84–99. [CrossRef]
40. Maertens, M.; Swinnen, J.F. Gender and modern supply chains in developing countries. *J. Dev. Stud.* **2012**, *48*, 1412–1430. [CrossRef]
41. Lei, M.; Lomax, R.G. The effect of varying degrees of nonnormality in structural equation modeling. *Struct. Equ. Model.* **2005**, *12*, 1–27. [CrossRef]
42. Abma, I.L.; Rovers, M.; van der Wees, P.J. Appraising convergent validity of patient-reported outcome measures in systematic reviews: Constructing hypotheses and interpreting outcomes. *BMC Res. Notes* **2016**, *9*, 226. [CrossRef] [PubMed]
43. Boomsma, A.; Hoogland, J.J. The robustness of LISREL modeling revisited. Structural Equation Models: Present and Future. *Festschr. Honor. Karl Jöreskog* **2001**, *2*, 139–168.
44. Browne, M.W.; Cudeck, R. Alternative ways of assessing model fit. In *Testing Structural Equation Models*; Bollen, K.A., Long, J.S., Eds.; Sage: Newbury Park, CA, USA, 1993; pp. 136–162.
45. Hu, L.; Bentler, P. Evaluating model fit. In *Structural Equation Modeling: Concepts, Issues, and Applications*; Hoyle, R.H., Ed.; Sage: London, UK, 1995; pp. 76–99.
46. Jöreskog, K.; Sörbom, D. *Structural Equation Modeling with the SIMPLIS*; Scientific Software International: Lincolnwood, IL, USA, 1993.
47. Bentler, P.M.; Bonett, D.G. Significance tests and goodness of fit in the analysis of covariance structures. *Psychol. Bull.* **1980**, *88*, 588. [CrossRef]

48. Bentler, P.M. Comparative fit indexes in structural models. *Psychol. Bull.* **1990**, *107*, 238. [CrossRef] [PubMed]
49. Marsh, H.; Grayson, D. Latent variable models of multitrait-multimethod data. In *Structural Equation Modeling: Concepts, Issues and Applications*; Hoyle, R.H., Ed.; Sage: Thousand Oaks, CA, USA, 1995; pp. 177–198.
50. Jöreskog, K.G.; Sörbom, D. *LISREL 8: User's Reference Guide*; Scientific Software International: Lincolnwood, IL, USA, 1996.
51. Abdollahzadeh, G.; Damalas, C.A.; Sharifzadeh, M.S.; Ahmadi-Gorgi, H. Selecting strategies for rice stem borer management using the Analytic Hierarchy Process (AHP). *Crop Prot.* **2016**, *84*, 27–36. [CrossRef]
52. Ducey, A.J.; Coovert, M.D. Predicting tablet computer use: An extended Technology Acceptance Model for physicians. *Health Policy Technol.* **2016**, *5*, 268–284. [CrossRef]
53. Ortiz, D.L.; Batuecas, E.; Orrego, C.E.; Rodríguez, L.J.; Camelin, E.; Fino, D. Sustainable management of peel waste in the small-scale orange juice industries: A Colombian case study. *J. Clean. Prod.* **2020**, *265*, 121587. [CrossRef]
54. Castro, L.; Escalante, H.; Jaimes-Estévez, J.; Díaz, L.J.; Vecino, K.; Rojas, G.; Mantilla, L. Low-cost digester monitoring under realistic conditions: Rural use of biogas and digestate quality. *Bioresour. Technol.* **2017**, *239*, 311–317. [CrossRef] [PubMed]
55. Lee, Y.H.; Hsiao, C.; Purnomo, S.H. An empirical examination of individual and system characteristics on enhancing e-learning acceptance. *Australas. J. Educ. Technol.* **2014**, *30*, 561–579. [CrossRef]
56. Wallace, L.G.; Sheetz, S.D. The adoption of software measures: A technology acceptance model (TAM) perspective. *Inf. Manag.* **2014**, *51*, 249–259. [CrossRef]
57. Abdullah, F.; Ward, R. Developing a General Extended Technology Acceptance Model for E-Learning (GETAMEL) by analysing commonly used external factors. *Comput. Hum. Behav.* **2016**, *56*, 238–256. [CrossRef]
58. Rezaei, R.; Safa, L.; Ganjkhanloo, M.M. Understanding farmers' ecological conservation behavior regarding the use of integrated pest management-an application of the technology acceptance model. *Glob. Ecol. Conserv.* **2020**, *22*, e00941. [CrossRef]

water

MDPI

Article

Life Cycle Assessment of the Mesophilic, Thermophilic, and Temperature-Phased Anaerobic Digestion of Sewage Sludge

Iryna Lanko [1,2,*], Laura Flores [1], Marianna Garfí [1], Vladimir Todt [3], John A. Posada [4], Pavel Jenicek [2] and Ivet Ferrer [1,*]

1 GEMMA—Group of Environmental Engineering and Microbiology, Department of Civil and Environmental Engineering, Universitat Politècnica de Catalunya·BarcelonaTech, c/Jordi Girona 1–3, Building D1, 08034 Barcelona, Spain; laura.flores.rosell@upc.edu (L.F.); marianna.garfi@upc.edu (M.G.)

2 Department of Water Technology and Environmental Engineering, University of Chemistry and Technology, Technicka, 5, 166 28 Prague, Czech Republic; jenicekp@vscht.cz

3 Veolia Česká Republika, a.s., Na Florenci 2116/15, Reception D, 110 00 Prague, Czech Republic; vladimir.todt@veolia.com

4 Department of Biotechnology, Delft University of Technology, van der Maasweg 9, 2629 HZ Delft, The Netherlands; J.A.PosadaDuque@tudelft.nl

* Correspondence: irynalanko@gmail.com (I.L.); ivet.ferrer@upc.edu (I.F.)

Received: 18 August 2020; Accepted: 3 November 2020; Published: 10 November 2020

Abstract: In this study the environmental impact of the anaerobic digestion (AD) of sewage sludge within an activated sludge wastewater treatment plant (WWTP) was investigated. Three alternative AD systems (mesophilic, thermophilic, and temperature-phased anaerobic digestion (TPAD)) were compared to determine which system may have the best environmental performance. Two life cycle assessments (LCA) were performed considering: (i) the whole WWTP (for a functional unit (FU) of 1 m^3 of treated wastewater), and (ii) the sludge line (SL) alone (for FU of 1 m^3 of produced methane). The data for the LCA were obtained from previous laboratory experimental work in combination with full-scale WWTP and literature. According to the results, the WWTP with TPAD outperforms those with mesophilic and thermophilic AD in most analyzed impact categories (i.e., Human toxicity, Ionizing radiation, Metal and Fossil depletion, Agricultural land occupation, Terrestrial acidification, Freshwater eutrophication, and Ozone depletion), except for Climate change where the WWTP with mesophilic AD performed better than with TPAD by 7%. In the case of the SL alone, the production of heat and electricity (here accounted for as avoided environmental impacts) led to credits in most of the analyzed impact categories except for Human toxicity where credits did not balance out the impacts caused by the wastewater treatment system. The best AD alternative was thermophilic concerning all environmental impact categories, besides Climate change and Human toxicity. Differences between both LCA results may be attributed to the FU.

Keywords: anaerobic digestion (AD); biogas; life cycle assessment (LCA); methane; waste activated sludge (WAS); wastewater treatment plant (WWTP)

1. Introduction

In conventional activated sludge wastewater treatment plants (WWTP), excess sludge is continuously formed at the biological reactor of the wastewater treatment line. WWTP operational expenses to handle the produced excess biological sludge, namely waste activated sludge (WAS), together with primary sludge may go up to 50% [1–3]. It has been a long time since AD was adopted as one of the most effective solutions of sewage sludge treatment in terms of sludge reduction, stabilization, and resource recovery [4–6]. Previously, sewage sludge was considered only as waste, and its disposal on a regular basis was quite challenging [7,8]. Population growth and natural resources exhaustion made it crucial to find a way out of this situation. Nowadays, sludge is considered as a source of substances that can be recovered and reused [9–11].

The zero-waste approach and circular economy paradigm change the angle at which nutrients, metals, organic matter, and other substances from WAS can be converted into valuable materials like biofuels and biofertilizers. Resource recovery processes are being widely adopted, even though recovery of certain substances, for instance, phosphorus still can be nonprofitable from the perspective of economics [12] or environmental impacts [13]. This is mostly because of the low concentration of valuable substances in the influent wastewater. However, as the world prices for phosphorus as for irreplaceable fertilizer are increasing, its recovery from sludge has become extremely important in a long-term perspective [14]. Currently, the attention is drawn to the reject water obtained after the AD process where the concentration of nutrients is significantly higher than in the influent wastewater [15]. However, the temperature regime and configuration of AD may significantly influence the nutrient concentration in the reject water and its volume [1,16,17].

AD is a biological process where organic matter is being biodegraded under anaerobic conditions, leading to the production of biogas, a gaseous biofuel mostly composed of methane, along with the digestate, which may be reused as biofertilizer. AD can be implemented with some variations in temperature (mesophilic (M), 35–40 °C and thermophilic (T), 55–70 °C) [18] and configuration (one- and two-stage reactors) [1,19]. Normally, the AD process consists of four main stages, namely hydrolysis, acidogenesis, acetogenesis and methanogenesis, which take place in the same reactor in the case of one-stage AD. Separate functioning of the first thermophilic (hydrolytic and acidogenic) and second mesophilic (acetogenic and methanogenic) stages in two-stage AD is accomplished to overcome the drawbacks of one-stage systems [20,21].

The most widely used anaerobic digester is a mesophilic one-stage, since its operation is known as the most simple and stable [22,23]. Nevertheless, the tendency changes towards more metabolically efficient and pathogenically safe digesters such as thermophilic one-stage or temperature-phased two-stage reactors [24]. Temperature phased anaerobic digestion (TPAD) implies a combination of thermophilic and mesophilic one-stage digesters and performs with better stability than thermophilic and higher organic matter degradation rate than mesophilic, which means increased energy efficiency and better control of process parameters [21,25].

Despite the increasing interest in thermophilic and TPAD systems' application, there are few studies comparing the environmental impact of full-scale WWTP with different AD systems [26] and none concerning the comparison of the environmental impact of mesophilic, thermophilic, and TPAD systems. The environmental impact assessment would allow for defining which AD system is the most beneficial in terms of environmental protection. Thus, the objective of this study is to evaluate and compare the environmental impacts of WWTP with different AD systems (M, T, TPAD) using the life cycle assessment (LCA) methodology.

One methodological challenge that leads to variability in LCA results for multiproduct systems, such as WWTPs, is the choice of a functional unit (FU) and its effect on the ecosystem. In this study, two FUs are selected in order to demonstrate a more comprehensive picture of environmental influence of AD implemented in the sludge line alone and at the whole WWTP [27].

The LCA approach is here used not only as a standard practice to estimate the environmental burden of the technological process [22,28,29] on a micro level to compare the three AD options, but also as an alternative to build up a regulatory planning system on a meso level to increase the efficiency of project-level decision-making and to provide advice on potential improvements for the sector's management, and also to ensure the realization of strategic environmental goals [30].

2. Materials and Methods

2.1. Anaerobic Digestion Systems

The LCA compared three different AD systems, namely, mesophilic, thermophilic, and TPAD. Data on the operation and performance of these systems treating the same sewage sludge was gathered from three lab-scale digesters which were run during five months [31]. The substrate was thickened WAS, disintegrated through centrifugation (total solids (TS) = 71.8 ± 3.4 g/L, volatile solids (VS) = 42.3 ± 4.1 g/L, chemical oxygen demand (COD) = 64.1 ± 4.1 g/L). The main features of the mesophilic, thermophilic, and TPAD lab-scale systems are shown in Table 1.

Table 1. Performance of the mesophilic, thermophilic and temperature-phased anaerobic digestion lab-scale systems. Values given as mean ± standard deviation.

	Process Variable (Units)	Thermophilic	Mesophilic	TPAD	
				1st Stage	2nd Stage
Operational conditions	Temperature (°C)	57 ± 1.5	38 ± 1.5	57 ± 1.5	38 ± 1.5
	HRT (days)	19	19	2	17
	V (L)	8.45	8.45	1.0	8.45
	OLR (g VS/L·day)	2.25	2.25	2.24	
Biogas production	Methane production rate (L CH_4/L·day)	7.8 ± 0.5	7.1 ± 0.6	2.5 ± 0.4	9.9 ± 0.2
	Methane yield (L CH_4/g COD)	0.17 ± 0.02	0.16 ± 0.02	0.05 ± 0.01	0.19 ± 0.02
	Methane content in biogas (% CH_4)	61.7 ± 3.1	66.1 ± 1.8	58.9 ± 5.3	70.9 ± 2.7
Removal efficiency	TS removal (%)	20.5 ± 2.4	18.5 ± 1.7	24.5 ± 2.2	
	VS removal (%)	27.6 ± 1.9	26.0 ± 2.6	35.5 ± 1.2	
Effluent characteristics	pH	7.4 ± 0.06	7.5 ± 0.02	6.7 ± 0.13	7.6 ± 0.02
	TS (g/L)	57.1 ± 2.4	58.5 ± 1.7	55.3 ± 2.2	
	VS (g/L)	30.7 ± 1.9	31.3 ± 2.6	28.6 ± 1.2	
	VS/TS (%)	53.8	53.5	51.7	
	COD (g/L)	43.2 ± 4.8	44.6 ± 3.7	48.8 ± 5.7	42.1 ± 1.2

2.2. Life Cycle Assessment

LCA is an analytical tool which allows to assess the environmental impacts of a product, technology, or process according to the "cradle-to-grave" approach. This time-tested technique allows not only to evaluate the potential environmental risks, but also define the life cycle stage and type of environmental impacts. The application of this method may help to improve the studied product, technology, or process by making its life cycle more friendly to the environment. LCA consists of four main steps according to ISO 14040 (2006), and ISO 14042 (2006): (i) goal and scope definition; (ii) inventory analysis; (iii) impact assessment; (iv) result interpretation.

2.2.1. Goal and Scope Definition

The goal of this study was to compare the potential environmental impacts of three types of AD processes: (i) mesophilic; (ii) thermophilic; (iii) TPAD systems.

To this end, two LCA cases with two different functional units (FU) were conducted: first for an activated sludge WWTP with the three different AD systems (here named as WWTP-LCA); second for the sludge line alone with the three different AD systems (here named as SL-LCA).

Since it has been reported in literature that the choice of the FU may change the overall balance of environmental impacts from harmful to beneficial and vice-versa [28], these two FUs were chosen considering the major outputs and functions of a WWTP. For the first case, the FU was 1 m^3 of treated wastewater, as the main function of the WWTP is to treat the wastewater stream. For the second case, the FU was 1 m^3 of produced methane, as one of the main functions in AD systems (also the secondary one within WWTP) is to produce energy out of the methane contained in the biogas. Methane was taken instead of biogas, as biogas might have a different content of methane depending on the AD system.

The two FUs were coupled with the system expansion approach (the alternative production of energy and fertilizers) and adopted to evaluate the environmental burden of AD at both scales, the whole WWTP (to assess the contribution of the sludge line to the whole WWTP) and the sludge line alone (to highlight the potential environmental benefits from methane production as an additional function of the system beyond the wastewater treatment). The application of two FUs would demonstrate a deeper, more comprehensive and more transparent picture of AD implementation at the sludge line and at the WWTP. Thus, applying the two FUs helps to present the environmental profile of each AD from different points of view [4,27].

For both LCAs, a period of one-year operation was considered, as it is a timescale that is long enough to assess the averaged operational parameters of any WWTP, including the fact that the construction part was not estimated in the impacts' analysis. The impacts of the construction phase were not accounted for, since the dimensioned WWTP is the same for all three scenarios and LCAs, and it would make the difference among three AD systems' exploitation less evident [4,32].

To sum up, the system's boundaries—for the first and second LCAs—consider the year-around operation of the whole WWTP with three different AD technologies (Figure 1a), and the sludge line alone with three different AD technologies (Figure 1b), respectively. Thus, three scenarios were considered in each LCA; namely mesophilic (M), thermophilic (T), and temperature-phased anaerobic digestion (TPAD).

2.2.2. Inventory Analysis

Inventory data for the WWTP-LCA and SL-LCA are summarized in Tables 2 and 3. Among the inputs and outputs, the flows of materials and energy resources, gaseous emissions, and solid wastes were considered. When possible, data from a full-scale facility was used, which was complemented by data gathered in the lab-scale set-up described previously (Section 2.1) [33].

Full-scale data were taken from a WWTP with thermophilic AD operated by Veolia Česká Republica, a.s. The full-scale operation was used to scale-up and validate the laboratory reactor results. The rest of the information for the mesophilic one-stage and TPAD systems was calculated based on different literature sources and benchmarking data of Veolia, a.s., as described below.

The input parameters included energy consumption, anti-bulking, anti-foaming and dewatering agent dosage, gaseous emissions from the disposed digestates, digestate amounts. Calculated parameters were energy consumption according to the literature [34], specific biogas production based on benchmarking data of Veolia, a.s., as well as anti-bulking, anti-foaming and dewatering agents' dosages.

The final digestate amount for each AD system was estimated in correspondence with the VS destruction observed in the lab-scale digesters (Table 1), which was consistent with other AD systems studied in the literature [19,35,36].

Since the lab-scale set-up treated WAS and the full-scale plant used sewage sludge with 32.5% WAS, the annual biogas production was recalculated for the thermophilic scenario based on the specific biogas production of the corresponding laboratory reactor.

The heat and electricity production was calculated based on the annual biogas production and technical equipment data (efficiency of combined heat and power unit) given by Veolia, a.s.

Annual gas emissions (CH_4 and N_2O) from the wastewater treatment process and disposed digestates were estimated according to the literature [37,38].

Background data (chemicals, avoided fertilizers—that are contained in the digestate and can be used in agriculture, transportation, wastewater treatment in a municipal wastewater treatment plant, solid wastes, energy provider) were taken from Ecoivent 3.1 database [33]. For all electricity requirements, the Czech electricity mix was considered since the full-scale WWTP is located in the Czech Republic [39] (which is composed of fossil 58.5%, nuclear 35.3%, solar 2.8%, hydro 2.7%, wind −0.7%).

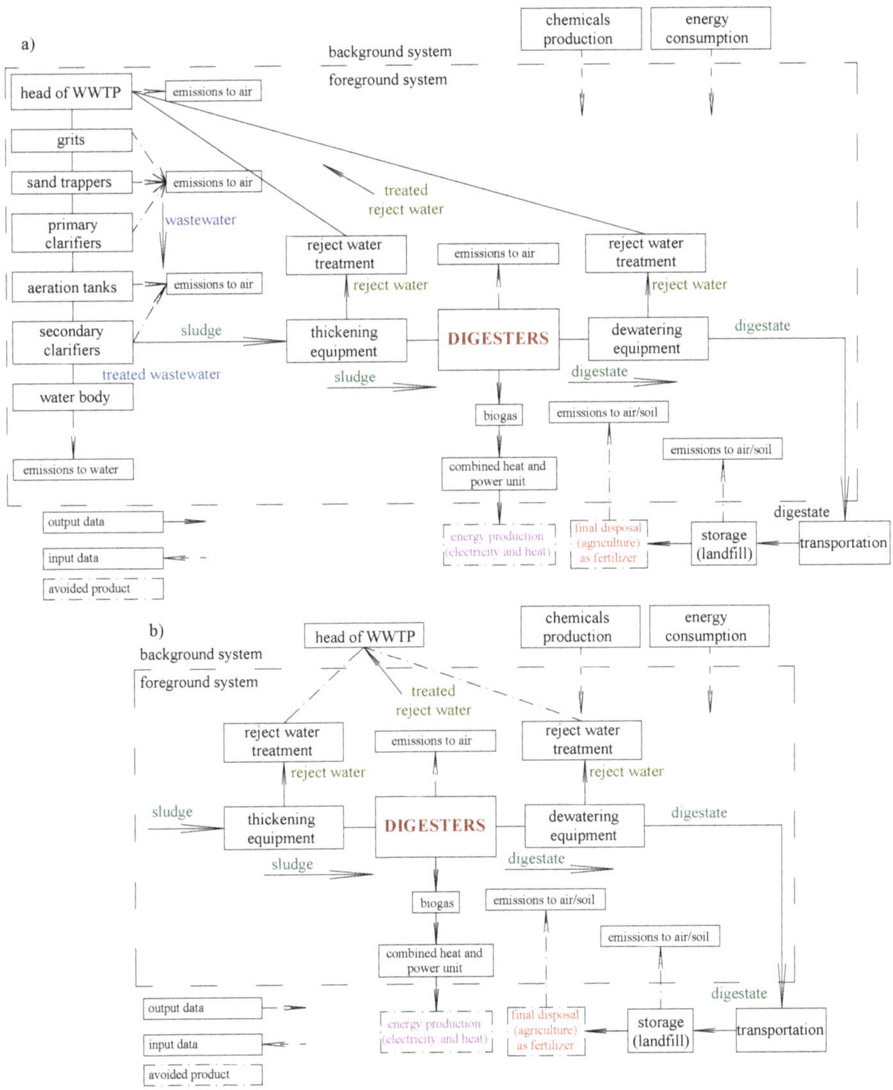

Figure 1. Flowcharts: (**a**) for life cycle assessment of the whole wastewater treatment plant (WWTP-LCA) and (**b**) for life cycle assessment of the sludge line alone (SL-LCA).

Table 2. Inventory data for the whole WWTP LCA (FU: 1 m^3 of treated wastewater).

Type of Data	WW/SL	Item	Referred to FU			Unit
			T	M	TPAD	
Input data						
Energy consumption	WW + SL	Total energy consumption	0.42	0.39	0.39	kWh/m^3
	WW	Only for WW	0.32	0.31	0.3	kWh/m^3
	SL	Only for SL	0.1	0.08	0.09	kWh/m^3
Reagents	WW	Phosphorus precipitation agent	0.07	0.07	0.03	kg/m^3
		Sludge anti-bulking agent	0.05	0.05	0.05	kg/m^3
	SL	Thickening agent	1.23×10^{-4}	1.23×10^{-4}	1.23×10^{-4}	kg/m^3
		Anti-foaming agent	2.63×10^{-4}	5.27×10^{-4}	5.27×10^{-4}	kg/m^3
		Dewatering agent	0.002	0.001	0.002	kg/m^3
		Chemo dezodoration agent	1.0×10^{-5}	1.0×10^{-5}	1.0×10^{-5}	kg/m^3
		Bio dezodoration agent	2.2×10^{-5}	2.2×10^{-5}	2.2×10^{-5}	kg/m^3
Output data						
Waste	WW	Sand	0.02	0.02	0.02	kg/m^3
		Sand	0.001	0.001	0.001	t*km/m^3
		Coarse waste	0.04	0.04	0.04	kg/m^3
		Coarse waste	0.002	0.002	0.002	t*km/m^3
Wastewater air emissions		CO	9	9	9	g/m^3
		SO_2	0.03	0.03	0.03	g/m^3
		Ozone	0.08	0.08	0.08	g/m^3
		N_2O	0.05	0.05	0.05	g/m^3
		H_2S	0.0008	0.0008	0.0008	g/m^3
		NH_3	2.8	2.8	2.8	g/m^3
		N_2O	1.8	1.81	1.81	g/m^3
		CO_2	81	81	81	g/m^3
		N_2	41	41	41	g/m^3
Wastewater contaminants		COD	0.04	0.04	0.04	kg/m^3
		SS	0.009	0.009	0.009	kg/m^3
		TN	0.02	0.02	0.02	kg/m^3
		TP	6.7×10^{-4}	6.7×10^{-4}	6.7×10^{-4}	kg/m^3
		As	9.6×10^{-6}	9.6×10^{-6}	9.6×10^{-6}	kg/m^3
		Pb	1.5×10^{-5}	1.5×10^{-5}	1.5×10^{-5}	kg/m^3
		Cd	1.2×10^{-6}	1.2×10^{-6}	1.2×10^{-6}	kg/m^3
		Cr	2.2×10^{-6}	2.2×10^{-6}	2.2×10^{-6}	kg/m^3
		Cu	2.9×10^{-5}	2.9×10^{-5}	2.9×10^{-5}	kg/m^3
		Ni	6.9×10^{-6}	6.9×10^{-6}	6.9×10^{-6}	kg/m^3
		Zn	3.2×10^{-5}	3.2×10^{-5}	3.2×10^{-5}	kg/m^3
		Hg	0.2×10^{-7}	0.2×10^{-7}	0.2×10^{-7}	kg/m^3
Digestate contaminants	SL	Dewatered digested sludge (wet)	0.82	0.86	0.78	kg/m^3
		Dewatered digested sludge (wet)	0.032	0.034	0.031	t*km/m^3
Avoided products		Electricity	0.23	0.22	0.35	kWh/m^3
		Heat	0.26	0.25	0.39	kWh/m^3
Air emissions		H_2S	2.0×10^{-4}	2.0×10^{-4}	2.0×10^{-4}	g/m^3
		CH_4	0.02	0.02	0.02	g/m^3
		CO_2, biogenic	28	28	28	g/m^3
		CO_2, not biogenic	44	44	44	g/m^3

Note: For T and TPAD, the transportation distance was 40 km to the sludge handling plant. According to the microbiological analyses and other sources [40], TPAD and T digestates meet the requirements of Class A biosolids. In case of M, due to pathogenic unsafety, the digestate was additionally treated by composting before its agricultural application. Therefore, it was transported over 2 × 40 km to the composting plant and back, so an additional energy consumption of 16 kWh/t of digestate for the purposes of composting was included [41].

Table 3. Inventory data for the sludge line LCA (FU: 1 m^3 of produced methane).

Type of Data	Item	Referred to FU			Unit
		T	M	TPAD	
Input data					
Energy consumption	AD energy consumption	1.32	1.12	0.79	kWh/m^3
	Energy consumption for composting	-	0.2	-	kWh/m^3
Reagents	Thickening agent	0.002	0.002	0.001	kg/m^3
	Anti-foaming agent	0.004	0.008	0.005	kg/m^3
	Dewatering agent	0.027	0.019	0.014	kg/m^3
	Chemo dezodoration agent	1.3×10^{-4}	1.4×10^{-4}	8.7×10^{-5}	kg/m^3
	Bio dezodoration agent	3.0×10^{-4}	3.1×10^{-4}	2.0×10^{-4}	kg/m^3
Output data					
Wastewater	Reject water	0.26	0.26	0.17	m^3/m^3
Wastes	Dewatered digested sludge (wet)	11.3	12.3	7.1	$kgDM/m^3$
	Dewatered digested sludge (wet)	0.45	0.49	0.28	$t*km/m^3$
Avoided products	Electricity	3.14	3.14	3.12	kWh/m^3
	Heat	3.51	3.51	3.49	kWh/m^3
Emissions	H_2S	0.003	0.003	0.002	kg/m^3
	CH_4	0.28	0.29	0.18	kg/m^3
	CO_2, biogenic	386	402	253	kg/m^3
	CO_2, not biogenic	607	633	398	kg/m^3

2.2.3. Impact Assessment

The LCA was performed with the software of *SimaPro® 8*. The potential environmental impacts were calculated by the ReCiPe midpoint method V1.12/Europe Recipe H [33].

Characterization was conducted for the following environmental impact categories: Climate change, Ozone depletion, Terrestrial acidification, Freshwater eutrophication, Human toxicity, Ionizing radiation, Agricultural land occupation, Metal depletion, and Fossil depletion [4]. The above mentioned nine environmental impact categories were selected and assessed considering their close connection with processes that take place in activated sludge WWTP with AD and that have been used in previous LCA studies [33,42].

Classification and characterization were performed as the only compulsory steps of impact assessment in terms of standards—ISO 14040 (2006) and ISO 14042 (2006).

2.2.4. Sensitivity Analysis

A sensitivity analysis on the digestate volume obtained from the TPAD system for both cases WWTP-LCA and SL-LCA was performed in order to take into account the influence that this parameter may have on the environmental impacts associated to digestate transport, treatment, and reuse.

The sensitivity analysis allowed to evaluate if and how the uncertainty of the assumed value in the inventory table could influence the final results. A variation of ±5% of the digestate volume was set for the TPAD scenario only, according to the variability of lab data obtained by the previous studies [19,26,35], and shortage of the reported data from full-scale WWTPs, since TPAD is the least spread AD system worldwide among others [43].

This analysis is done through the sensitivity coefficient (S) which indicates the sensitivity of a particular model output to the changes in the variable being considered. The S is calculated according to the following equation [44]:

$$\text{Sensitivity coefficient (S)} = \left(\frac{\text{Outputhigh} - \text{Outputlow}}{\text{Outputbaseline}}\right) / \left(\frac{\text{Inputhigh} - \text{Inputlow}}{\text{Inputbaseline}}\right) \quad (1)$$

where Input is the value of the input variable (i.e., digestate amount), and Output is the value of indicator according to the correspondent impact category.

3. Results and Discussion

3.1. Life Cycle Assessment

3.1.1. WWTP-LCA with Mesophilic, Thermophilic, or Temperature-Phased Anaerobic Digestion

The results of the LCA for the whole WWTP (WWTP-LCA) are shown in the Figure 2. This figure includes all environmental impact categories considered in this study, and within each impact category there are three scenarios: mesophilic, thermophilic, and TPAD. For each scenario, results are shown for the whole WWTP, and separately for the wastewater treatment line and the sludge treatment line; this disaggregation of results was done to better identify the contributions of each process stage to the overall impacts. Positive values represent the environmental impacts, while negative values refer to the avoided environmental impacts.

According to the results (Figure 2), the differences among the three AD scenarios are not significantly large, however there are some trends that are discussed here. First, the wastewater treatment line would lead to larger environmental impacts than those cases where the sludge line is incorporated into the AD system. Thus, any implemented AD improves the environmental status of the WWTP mainly due to the credits obtained from the substitution of electricity generation from the fossil fuels [45]. Similar results have been reported for other LCA studies on full-scale AD plants [4,26].

In general terms, TPAD has the lowest environmental impacts, in comparison to T and M, for eight out of the nine impact categories presented in Figure 2, except for Climate change. Furthermore, a one-to-one comparison between T and M shows that their calculated environmental impacts are virtually the same for five out the nine compared categories (i.e., Ionizing radiation, Agricultural land occupation, Metal depletion, Fossil depletion, and Freshwater eutrophication), and with a slightly better environmental performance (meaning lower environmental impacts) for T over M in two impact categories (i.e., Terrestrial acidification and Ozone depletion). T outperforms both M and TPAD in one category (i.e., Climate change), and has a slightly better performance than M in only one category (i.e., Human toxicity).

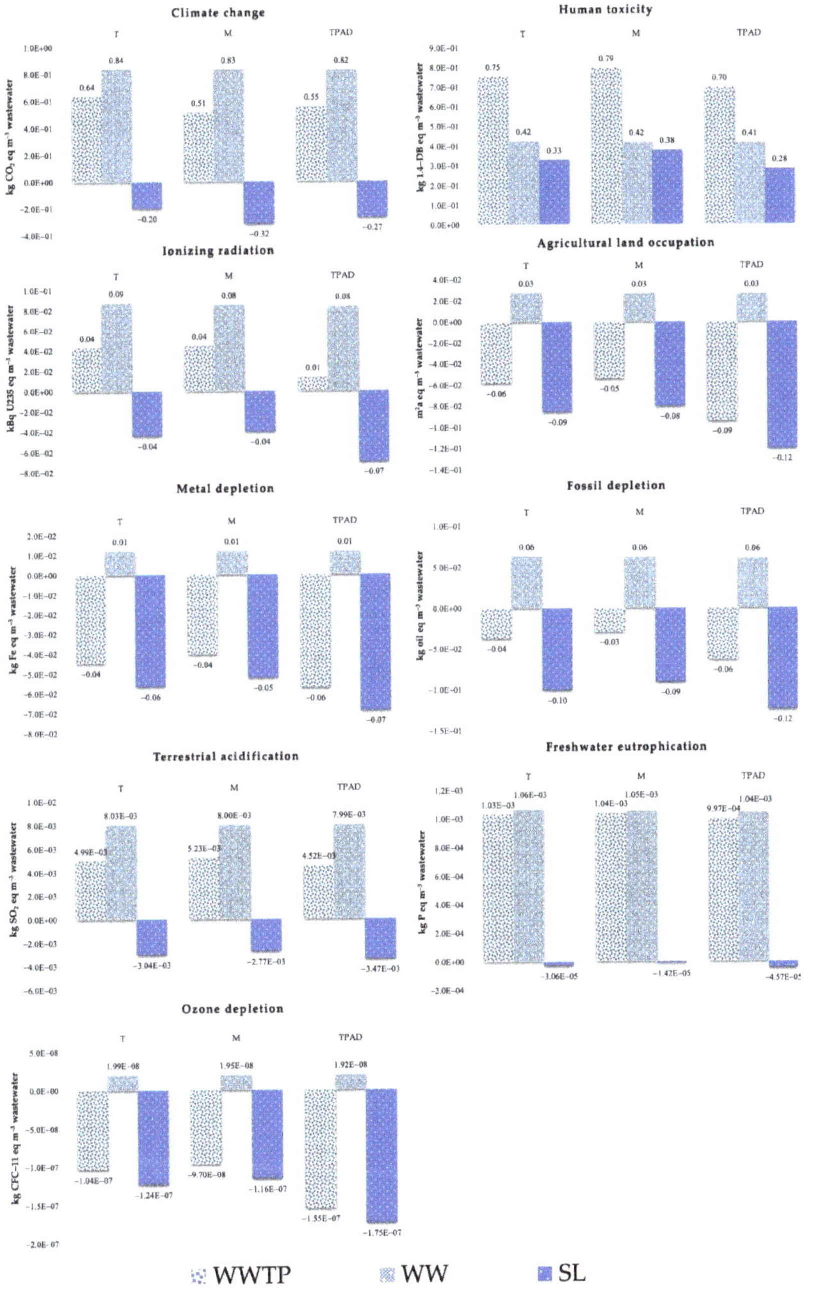

Figure 2. Potential environmental impacts for the three scenarios of the whole WWTP (WWTP-LCA): Mesophilic (M), Thermophilic (T), and Temperature-Phased Anaerobic Digestion (TPAD). WWTP: wastewater treatment plant; WW: wastewater line; SL: sludge line. Results shown for the FU: 1 m^3 of treated wastewater.

In the case of Climate change, the biggest impacts are caused by the wastewater treatment line (Figure 2). Conversely, the sludge line decreases the Climate change impacts up to 38% for M, around 35% for TPAD, and 24% for T. Climate change is related to nonrenewable energy consumption, which is especially high in the biological reactor of activated sludge WWTP, accounting for more than 50% of the total energy consumption—Table 2 and [46]. The positive influence of sludge line mainly comes from AD which supplies with the fertilizer obtained after WAS is digested (Sludge disposal_SL) that substituted the industrial production of the fertilizer with its harmful effect through Climate change. Additionally, AD generates renewable energy out of the biogas produced as a result of the organic matter biodegradation, and counterbalances nonrenewable energy consumption that would otherwise be required to fuel the process. The highest avoided impacts on Climate change are obtained with the mesophilic digestion (M) which is 40% and 42% better than T and TPAD, respectively. These avoided impacts occur due to a type of digestate disposal (which is composting and consequent agricultural land application for M, agricultural land application alone—for T and TPAD) and to its larger amount in comparison to T and TPAD (Table 1). In terms of the energy balance, TPAD is more beneficial than T and M by more than 50% within the sludge line. The lowest total avoided impact on Climate change is obtained with the thermophilic digestion, as a consequence of the energy balance of the process, i.e., energy produced vs. energy consumed by each AD system—Figure 3.

The results of WWTP-LCA regarding all constituents are depicted in Figure 3.

With regard to the factors that contributed the most to the environmental impact of Climate change, those emissions to the air (Air emissions_WW) and from the energy demand (Electricity_WW) in the wastewater treatment line are the most significant ones (i.e., around 60% and 25%, respectively, of all contributors from the wastewater treatment line). The contribution from the emissions to the air in the sludge line (Air emissions_SL) are only 5% (Figure 3). Hence, the total environmental impact was partly compensated by land application as the final sludge disposal (Sludge disposal_SL) and energy produced from the methane (Electricity_SL) obtained during AD with the following percentage of these two factor contributions, respectively: 72% and 28% for T, 82% and 18% for M, and 54% and 46% for TPAD. The balance of these two factors for TPAD shows better long-term performance of this AD technology.

For Human toxicity (Figure 2) the wastewater line constituents are quite similar in all scenarios, however, the absolute value of the sludge line varies: the larger negative effect to the environment is for M, 0.377 kg 1.4-DB eq, and the smaller one is for TPAD, 0.282 kg 1.4-DB eq, which is almost 25% less than that of M, and 15% less than T. This happens due to the higher total amount of digestate produced at M conditions rather than at T or TPAD. In particular for the Human toxicity category, less than 3% of the avoided environmental impacts are given by the energy production at TPAD conditions. The main contributor to this impact category is land application (Sludge disposal_SL) due to the heavy metals and other toxic substances that are still present in the digestate (41–45%)—Figure 3. The other major contributors are the energy consumption in the wastewater treatment line (26–29%), followed by the water body pollution (Water pollution_WW) made by treated wastewater discharge (19–22%) and, finally, by the different chemicals' consumption (Chemicals_WW) used at different stages of the wastewater treatment processes such as phosphorus precipitation and coagulation (all around 5%).

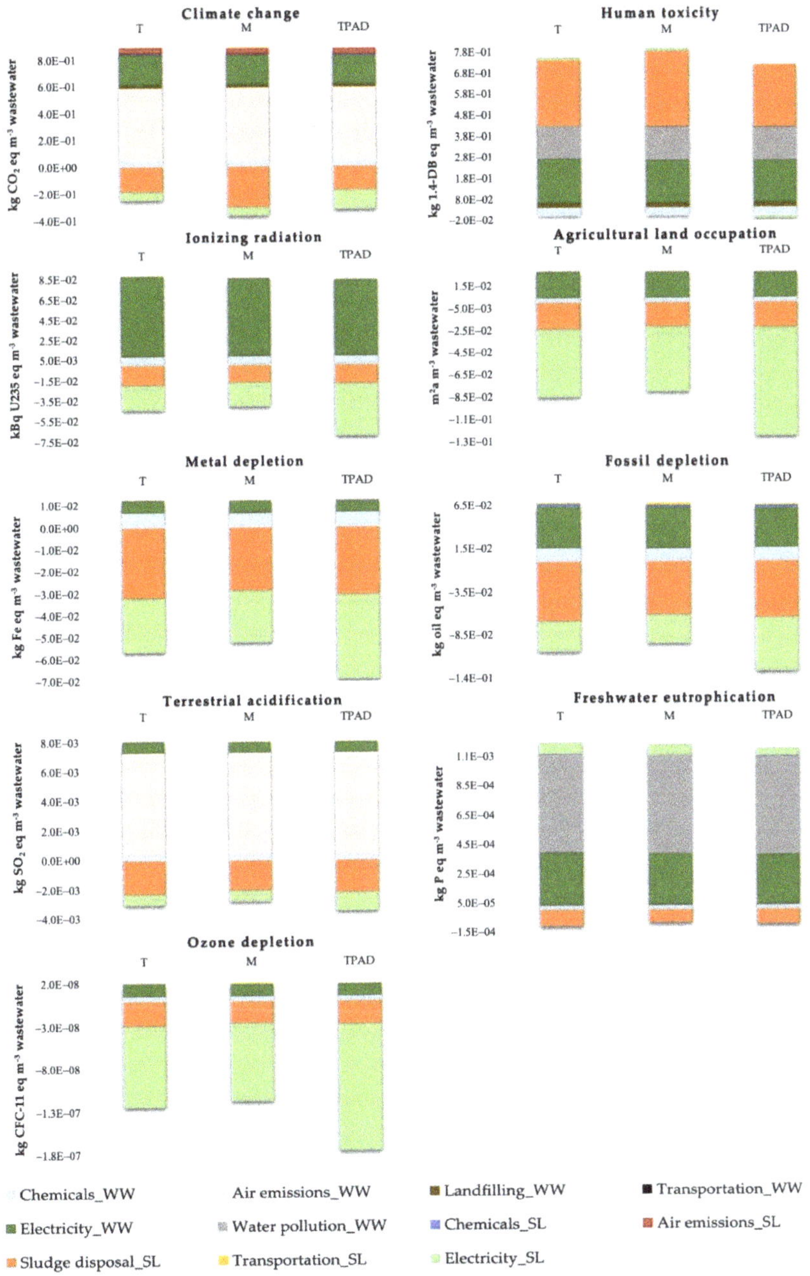

Figure 3. Contribution analysis of the potential environmental impacts for the three scenarios of both wastewater and sludge lines (WWTP-LCA): Mesophilic (M), Thermophilic (T), and Temperature-Phased Anaerobic Digestion (TPAD). Results shown for the FU: 1 m^3 of wastewater treated.

In terms of the Ionizing radiation impact category, even though the absolute values for both lines are lower than ±0.1 kBq U235 eq/m^3 of treated wastewater, the avoided environmental impacts given by the sludge line of WWTP compensates the negative influence of wastewater treatment line for more than 40% for both T and M, and around 90% for TPAD (Figure 2). The latter leads to a better balance of both avoided and overall environmental impacts for TPAD. The rest of the contributions are given by different chemicals' consumption (Chemicals_WW) used for wastewater treatment processes such as phosphorus precipitation and coagulation (all less than 9%)—Figure 3. The factors that represent the avoided environmental impact are land application as the final sludge disposal (Sludge disposal_SL) and the energy production (Electricity_SL), both are from the sludge line of WWTP. The percentage contributions of them, respectively, are 45% and 55% for T, 43% and 57% for M, 27% and 73% for TPAD. For TPAD, the distribution is significantly different from T and M due to the better energy balance after AD.

In the context of other impact categories such as Agricultural land occupation, and Metal and Fossil depletion, the avoided impacts of the sludge line made mainly by land application as the final sludge disposal (Sludge disposal_SL) and energy production (Electricity_SL) completely captures the negative influence of wastewater treatment line given by energy consumption (Electricity_WW) and chemicals' consumption (Chemicals_WW)—Figure 3.

The environmental impact represented through the rest of the assessed impact categories show relatively low absolute values: <0.0052 kg SO_2 eq/m^3 of treated wastewater for Terrestrial acidification, <0.0011 kg P eq/m^3 of treated wastewater for Freshwater acidification, and $<-1.0 \times 10^{-7}$ kg CFC-11 eq/m^3 of treated wastewater for Ozone depletion.

Terrestrial acidification impacts are built up due to the gaseous emissions (Air emissions_WW) from the wastewater treatment line, <10% from energy demand (Electricity), and <5% from chemicals (Chemicals_WW) used at wastewater treatment line—Figure 3.

Freshwater eutrophication is mostly affected by water body pollution (Water pollution_WW) with a 45–50% contribution—Figure 3—and by energy consumption (Electricity_WW) with a 25% contribution from the wastewater treatment line.

Ozone depletion results are driven by the avoided environmental impacts of both the energy produced (Electricity_SL)—around 60% for T and M, and more than 65% for TPAD; and the land application (Sludge disposal_SL)—around 20% for T and M, and around 15% for TPAD (see Figure 3). These avoided impacts are significantly bigger than those caused by Electricity_WW and Chemicals_WW consumption.

Concerning the factors mainly contributing to the different environmental impact categories negatively, there are certain ones confirming their prevailing parts in the total environmental burden. In the case of WWTP-LCA, the major contributors are the gaseous emissions from the open biological step reservoirs to the air, the energy consumption for aeration tanks [47], and the water body secondary pollution given by treated wastewater discharge—see Figure 3. All of them are related to the wastewater treatment line.

For the Climate change impact category, both the gaseous emissions to the air and the energy consumption—again related to the wastewater treatment line—are the biggest contributors to the environmental burden, followed by Chemicals consumption—related to the wastewater treatment line—and the gaseous emissions to air—from the sludge line—with around 10–15% all together.

3.1.2. SL-LCA with Mesophilic, Thermophilic, or Temperature-Phased Anaerobic Digestion

The LCA results of the sludge line (SL-LCA) including methane production are presented in Figure 4 using the second FU: 1 m^3 of methane produced (unlike Figures 2 and 3 which use the FU: 1 m^3 of wastewater treated). Figure 4 includes all environmental impact categories considered in this study, and within each impact category there are three scenarios: mesophilic, thermophilic, and TPAD. Furthermore, the different contributions from all process inputs and outputs are included for each of the three scenarios and for all categories. Positive values represent the environmental impacts, while negative values refer to the avoided environmental impacts (here considered as environmental credits).

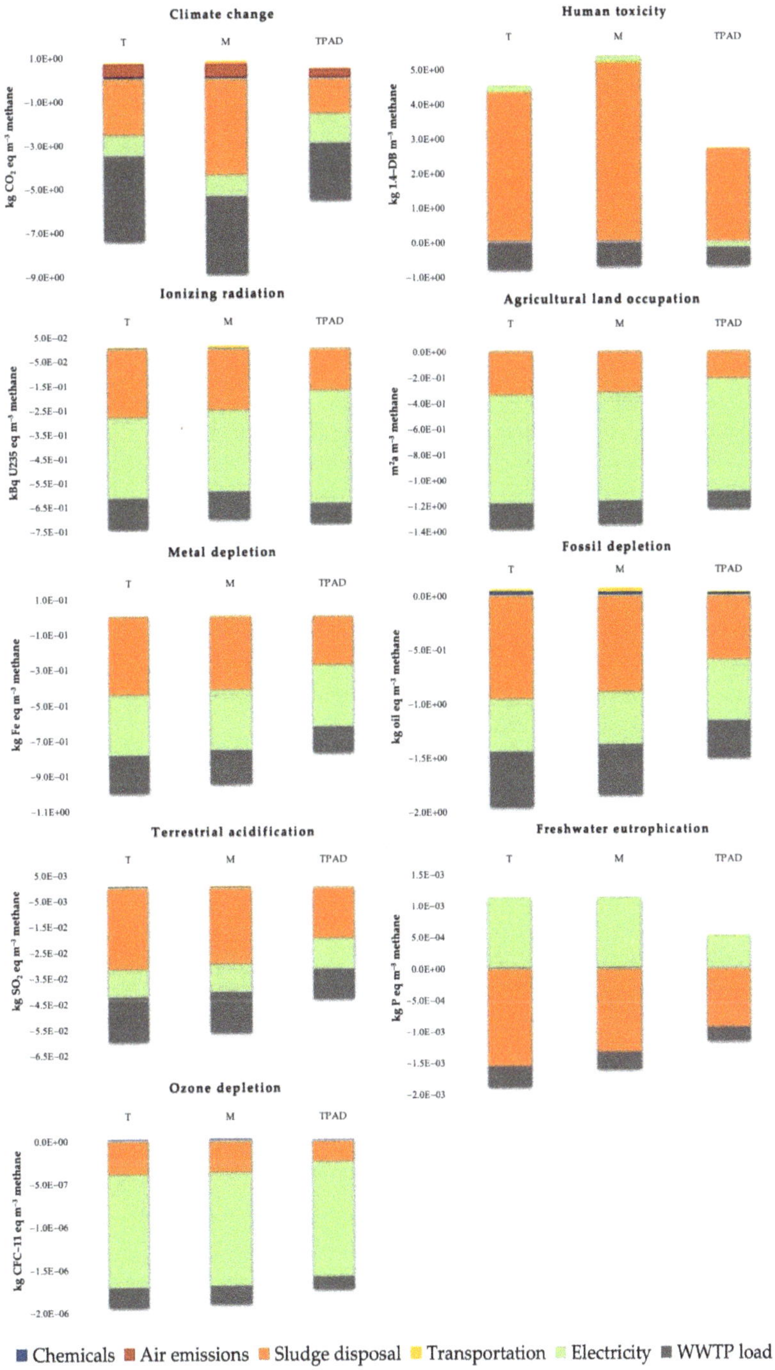

Figure 4. Contribution analysis of the potential environmental impacts for the three scenarios of the sludge line (SL-LCA): Mesophilic (M), Thermophilic (T), and Temperature-Phased Anaerobic Digestion (TPAD). Results shown for the FU: 1 m^3 of methane produced.

The aggregated SL-LCA results for the three scenarios lead to overall avoided environmental impacts in all categories with exception of Human toxicity. From the three scenarios, T outperforms M and TPAD in seven out of the nine impact categories here analyzed (except for Climate change and Human toxicity) (Figure 4). M is consistently the second best scenario in six out of the nine categories, except for Climate change (where it performs the best), and both Ionizing radiation and Human health (where it performs worse). Finally, TPAD has the lowest environmental impacts for Human health (by over 50%), while it also has the least avoided environmental impacts for seven categories out of the nine here analyzed (Figure 4).

In all SL-LCA scenarios, the contributing factors with the largest absolute values (i.e., either potential impacts (for Human toxicity) or avoided impacts (for all the other categories)) are the final sludge disposal (starting from 15% for Ozone depletion to almost 80% for Human toxicity), energy balance (from 12% for Climate change to over 75% for Ozone depletion), and water pollution (from 11% for Human toxicity and Ionizing radiation to 22% for Fossil depletion). It is also important to highlight that the factor of gaseous emissions to the air contributes significantly in a harmful way only for Climate change (more than 80% of caused environmental impact and only less than 8% of total environmental impact) due to the digestate accumulated at the landfill [26].

In the case of specific impact categories, the avoided environmental impacts in Climate change for both T and M are larger than those of TPAD by 24% and 38%, respectively—Figure 4. The only factor causing environmental impacts on Climate change for the three scenarios are the gaseous emissions from AD installations (Air emissions). On the contrary, the avoided environmental impacts have been credited by the following factors: additional reject water treatment (WWTP load concerning each scenario: 53% for T, 40% for M, and 47% for TPAD), final sludge disposal (Sludge disposal: 34% for T, 49% for M, and 29% for TPAD) and energy production (Electricity: 13% for T, 11% for M, and 24% for TPAD). Further minor avoided impacts are related to chemical consumption (Chemicals) (around 10% for all AD types) and transportation (Transportation) (around 5% for T and TPAD, and around 11% for M due to longer distance—a round trip—to the composting site).

In terms of Human toxicity, TPAD demonstrates the best results with the lowest environmental impacts at SL-LCA (Figure 4). The TPAD's impacts on Human toxicity are 46% lower than T, and 58% lower than M. The most substantial contribution to Human toxicity is coming from the final sludge disposal (Sludge disposal—95% for T, 96% for M, and 99% for TPAD) which makes sense as it is the agricultural land application for the T and TPAD scenarios and agricultural land application via composting for M scenario—Figure 4 and Table 3. The rest of the impacts on Human toxicity are mostly caused by the energy consumption (Electricity) with 4% and 3%, for T and M, respectively. While for TPAD, the energy balance is slightly positive, meaning that the system produces surplus energy with respect to its total consumption which leads to avoided impacts by almost 6%. Hence, the TPAD scenario for SL can be considered as energy self-sufficient process and an electricity supplier. Furthermore, for Human toxicity there are some minor avoided impacts from additional reject water treatment (WWTP load—100% for T and M and 94% for TPAD) which is highly polluted, meaning that it can be used as an additional source for resource recovery [15,48].

Interestingly, Human toxicity is the only impact category that does not result in overall avoided impacts at the sludge line. This happens due to the sufficient amounts of heavy metals and other toxic pollutants that are not completely removed during AD operation. Knowing that, T's and TPAD's digestates are considered to be pathogenically safe, and their final disposal can be a direct land application as fertilizers [35]. M digestate undergoes an additional step of composting prior to its application in agriculture. However, the gaseous emissions as well as the traces of heavy metals (Table 3) result in certain danger to the human health [9].

Looking at the Freshwater eutrophication impact category, the TPAD scenario shows both the lowest environmental impacts (50% lower than T and M, with energy consumption—Electricity—as the main contributor) and the lowest avoided environmental impacts. In the latter case, the prevailing contributors are digestate usage for the agricultural land application (Sludge disposal) and water body pollution reduction (WWTP load).

The rest of the impact categories (i.e., Ionizing radiation, Agricultural land occupation, Metal and Fossil depletion, Terrestrial acidification, Freshwater eutrophication, and Ozone depletion) follow a similar pattern. For all SL-LCA scenarios, the overall result can be referred as avoided impacts, with the best results being obtained for T, followed by M, and finally by TPAD. The main contributors are Sludge disposal and Electricity, and the WWTP load to a lower extent (with a maximum of 20% for Terrestrial acidification and lesser for other impact categories).

3.2. Sensitivity Analysis

The sensitivity response (i.e., the sensitivity coefficient "S" as described in Section 2.2.4) of all studied environmental impact categories was analyzed with respect to the assumed values for the digestate volume (with ±5% of the baseline value for the TPAD scenario, i.e., 90,332 t/year > 85,816 t/year > 81,299 t/year). Only the TPAD scenario was considered for sensitivity analysis due to the variability of the experimental data obtained by the previous studies [19,26,35], and shortage of the reported data from full-scale WWTPs, especially considering that TPAD is the least spread AD system worldwide among others [43].

The sensitivity coefficients were analyzed considering the processing conditions of TPAD for both the WWTP and the SL alone as shown in Table 4.

Table 4. Sensitivity coefficients (S) and environmental impacts of the whole WWTP (WWTP-LCA) and SL (SL-LCA) with respect to the TPAD baseline value assumed for the digestate volume.

Case	WWTP			SL		
Impact Category	S Coefficient	+5%	−5%	S Coefficient	+5%	−5%
Climate change, kg CO_2 eq/FU	−0.309	0.543	0.561	0.321	−5.108	−4.938
Human toxicity, kg 1.4-DB eq/FU	0.431	0.711	0.679	**1.325**	**2.162**	**1.880**
Ionising radiation, kBq U235 eq/FU	−1.424	0.012	0.014	0.241	−0.722	−0.704
Agricultural land occupation, m^2a/FU	0.245	−0.096	−0.093	0.173	−1.226	−1.204
Metal depletion, kg Fe eq/FU	0.530	−0.059	−0.056	0.363	−0.777	−0.748
Fossil depletion, kg oil eq/FU	0.980	−0.068	−0.061	0.415	−1.495	−1.431
Terrestrial acidification, kg SO_2 eq/FU	−0.478	4.41×10^{-3}	4.64×10^{-3}	0.467	−0.044	−0.042
Freshwater eutrophication, kg 1.4-DB eq/FU	−0.103	-9.92×10^{-4}	-1.0×10^{-3}	**1.504**	$\mathbf{-6.76 \times 10^{-4}}$	$\mathbf{-5.77 \times 10^{-4}}$
Ozone depletion, kg CFC-11 eq/FU	0.172	-1.57×10^{-7}	-1.54×10^{-7}	0.143	-1.73×10^{-6}	-1.71×10^{-6}

Note: Sensitivity coefficients (S) are unitless; Units of each environmental impact category consider the specific FU for WWTP and SL, i.e., 1 m^3 of treated wastewater and 1 m^3 of produced methane, respectively. Bold numbers are for the most sensitive impact categories (with S > 1.0).

A positive value of the sensitivity coefficient (S) refers to a straight influence of the studied parameter on the environmental results: e.g., the more sludge that is considered, the higher the (avoided) environmental impacts are. On the contrary, a negative sensitivity coefficient means an opposite influence of the studied parameter on the environmental results, it is e.g., the more sludge that is considered, the less the (avoided) environmental impacts are.

In this study, negative sensitivity coefficients are obtained only for WWTP-LCA, concerning four impact categories: Climate change, Ionizing radiation, Terrestrial acidification, and Freshwater eutrophication.

In a case of the potential environmental impacts related to the Climate change, Ionizing radiation, and Terrestrial acidification (Table 4), this behavior occurs due to an increased (proportional to the digestate volume) amount of both digestate as fertilizer substituent and energy recovered as biogas. In the case of Freshwater eutrophication, this opposite behavior occurs since an increase in the digestate volume leads to an additional amount of highly polluted reject water (that in turn needs to be further treated) generating a minor amount additional environmental impacts but that overall reduces the total avoided impacts.

On the contrary, the sensitivity coefficients for SL-LCA are positive values in all impact categories indicating a positive relation between the input variable (i.e., the assumed digestate volume) and the out variable (each environmental impact category). In this case, larger digestate volumes lead to the larger (either potentially caused or avoided) environmental impacts. In particular for the Human toxicity category, impacts are higher with the increase in the digestate volume (due to the proportional increase in the present pollutants in the digestate), while all the other categories included in Table 4 result in larger avoided environmental impacts with the increment in the digestate volume (due to the production of the avoided products such as fertilizers and electricity).

The most sensitive environmental impact categories in terms of ±5% variation in the digestate amount are, at SL-LCA, Human toxicity and Freshwater eutrophication, as S is positive and higher than 1.0.

The digestate amount variation of +5% increases the environmental burden of Human toxicity by 6.5% from the baseline value at SL-LCA. At WWTP-LCA, the digestate amount of ±5% had the influence of around 2.0% referring to the baseline value. It is also important to mention that the Human toxicity impact category is the only one with positive sensitivity coefficients at both LCAs, and for SL-LCA the sensitivity coefficient is higher than 1.0. Hence, it is important to mention that such sensitivity behavior of the Human toxicity category reveals that the major environmental concern based the variability of the digestate amounts would be on this impact category.

For Freshwater eutrophication at SL-LCA, the avoided environmental impacts increase by over 7.0% along with the increment in the digestate amount applied in agriculture as a fertilizer. At WWTP-LCA, the sensitivity coefficient at this impact category is negative and lower than 1.0, and can be neglected.

At the WWTP-LCA, the rest of the impact categories (i.e., apart from the ones with negative sensitivity coefficient) result in S values lower than 1.0. The sensitivity coefficient values higher than 0.5 are for the impact categories Metal and Fossil depletion. These two impact categories are affected by 2.5% to 5%, respectively (Table 4), and they refer to overall avoided environmental impacts. Furthermore, the sensitivity coefficients for Metal and Fossil depletion at WWTP-LCA are higher than those at SL-LCA. A reason for such a difference is the contribution in the energy balance (Electricity_WW) at WW line (Tables 3 and 4). The absolute values of avoided environmental impacts at ±5% of digestate are essentially higher at SL-LCA than at WWTP-LCA for Metal depletion (on over 92%) and for Fossil depletion (on over 95%) due to the energy consumption at the WW line concerning both impact categories.

In general terms, it can be said that the WW line has a higher harmful effect on the environment than SL line itself, and the larger its scale is, the larger the potential environmental impacts will be, contrary to the SL line.

Other general trends from the sensitivity analysis are that the sensitivity gives a clear overview that AD, namely TPAD, affects the environment mainly due to the toxic substances' content and air emissions derived from the digestate, which are proportional to its volume. The digestate production affects the environment negatively by the contribution to Human toxicity due to the final sludge disposal (Sludge disposal) coming from the SL line which relates to both WWTP-LCA and SL-LCA (Figures 3 and 4). At the same time, digestate production has also a positive effect given by resource (fertilizer) and energy (electricity and heat) recovery (Sludge disposal and Electricity, respectively) and also due to the additional reject water treatment (WWTP load) derived from the SL line (Figure 4).

Therefore, the impact categories of Human toxicity, Metal and Fossil depletion which are directly related to the produced digestate amount are of major attention for these types of processes. Considering the case of the TPAD scenario, it can also be said that ±5% of digestate production does not affect most of the (avoided) environmental impacts. Only three environmental impact categories have S > 1.0, namely: Human toxicity (SL-LCA), Ionizing radiation (WWTP-LCA), and Freshwater eutrophication (SL-LCA). These S values, bigger than 1.0, are strongly related to several contributing factors such as energy consumption (WWTP-LCA), final sludge disposal, and reject water treatment (SL-LCA)—Figures 3 and 4, and Table 4. Hence, these findings of the sensitivity analysis should be considered and taken into account for future designs of WWTPs and AD systems.

Based on the sensitivity analysis performed, it can be said that the main factor that contributes to the environmental impact through Human toxicity impact category is digestate quality (pathogenic safety, presence of the toxic substances, and gaseous emissions) and its amount. Therefore, by considering and changing the final digestate disposal, the total environmental impact can be reduced.

4. Conclusions

In this study, a comparative LCA analysis was carried out to evaluate the environmental impacts of three alternative AD processes (mesophilic, thermophilic, and TPAD) used for sludge treatment in activated sludge WWTP. The environmental burden was evaluated at two scales, namely the whole WWTP (to assess the contribution of the sludge line to the whole WWTP)—with a FU of 1 m^3 treated wastewater—and the sludge line alone (to highlight the potential environmental benefits from methane production as an additional function of the system beyond the wastewater treatment)—with a FU of 1 m^3 produced methane.

In the WWTP-LCA, five (Climate change, Human toxicity, Ionizing radiation, Terrestrial acidification, and Freshwater eutrophication) out of the nine environmental impact categories analyzed showed potential environmental impacts. The rest of the environmental impact categories (Agricultural land occupation, Metal and Fossil depletion, Ozone depletion) showed avoided environmental impacts, since the WW line led to potential environmental impacts in all impact categories, while the SL line led to avoided environmental impacts for most environmental impact categories (except for Human toxicity). Among all scenarios, the WWTP with TPAD outperformed those with mesophilic and thermophilic AD in all the environmental impact categories, besides Climate change.

The SL-LCA showed mostly avoided impacts, being the highest for thermophilic AD, followed by mesophilic AD and TPAD, except for Climate change where mesophilic AD was the most beneficial. The only potential environmental impact was Human toxicity, being the lowest for TPAD.

Differences between both LCA results may be attributed to the FU.

In addition, it can be also concluded that such products as nutrients and energy recovered from the AD systems and incorporated into the sludge treatment create an amount of credits that make the whole WWTP more environmentally friendly.

Author Contributions: Conceptualization, P.J., I.F. and M.G.; Methodology, J.A.P., M.G. and I.F.; software, I.L. and L.F.; validation, M.G., V.T., P.J. and I.F.; formal analysis, I.L.; investigation, I.L.; resources, V.T. and P.J.; data curation, I.L. and L.F.; Writing—original draft preparation, I.L.; Writing—review and editing, J.A.P., I.F. and P.J.; supervision, P.J. and I.F.; project administration, P.J.; funding acquisition, P.J. and I.F. All authors have read and agreed to the published version of the manuscript.

Funding: The research was funded by the European Union's Horizon 2020 research and innovation programme under the Marie Skłodowska-Curie grant agreement No 676070. This communication reflects only the authors' views and the Research Executive Agency of the EU is not responsible for any use that may be made regarding the information it contains.

Acknowledgments: I.F. and M.G. are grateful to the Government of Catalonia (Consolidated Research Group 2017 SGR 1029). M.G. acknowledges the Spanish Ministry of Economy and Competitiveness (RYC-2016-20059).

Conflicts of Interest: The authors declare no conflict of interest. The funders had no role in the design of the study; in the collection, analyses, or interpretation of data; in the writing of the manuscript, or in the decision to publish the results.

Abbreviations

AD	anaerobic digestion
COD	chemical oxygen demand
HRT	hydraulic retention time
LCA	life cycle assessment
M	mesophilic
OLR	organic loading rate
SL	sludge treatment line
SL-LCA	LCA performed within the boundaries of the sludge line
T	thermophilic
TPAD	temperature-phased anaerobic digestion
TS	total solids
V	reactor working volume
VS	volatile solids
WAS	waste activated sludge
WW	wastewater treatment line
WWTP	wastewater treatment plant
WWTP-LCA	LCA performed within the boundaries of the whole wastewater treatment plant

References

1. Leite, W.R.M.; Gottardo, M.; Pavan, P.; Filho, P.B.; Bolzonella, D. Performance and energy aspects of single and two phase thermophilic anaerobic digestion of waste activated sludge. *Renew. Energy* **2016**, *86*, 1324–1331. [CrossRef]
2. Yang, P.; Li, D.; Zhang, W.; Wang, N.; Yang, Z.; Wang, D.; Ma, T. Flocculation-dewatering behavior of waste activated sludge particles under chemical conditioning with inorganic polymer flocculant: Effects of typical sludge properties. *Chemosphere* **2019**, *218*, 930–940. [CrossRef]
3. Wei, H.; Gao, B.; Ren, J.; Li, A.; Yang, H. Coagulation/flocculation in dewatering of sludge: A review. *Water Res.* **2018**, *143*, 608–631. [CrossRef]
4. Lamnatou, C.; Nicolaï, R.; Chemisana, D.; Cristofari, C.; Cancellieri, D. Biogas production by means of an anaerobic-digestion plant in France: LCA of greenhouse-gas emissions and other environmental indicators. *Sci. Total. Environ.* **2019**, *670*, 1226–1239. [CrossRef]
5. Rajendran, K.; Murthy, G.S. Techno-economic and life cycle assessments of anaerobic digestion—A review. *Biocatal. Agric. Biotechnol.* **2019**, *20*, 101207. [CrossRef]
6. Wainaina, S.; Awasthi, M.K.; Sarsaiya, S.; Chen, H.; Singh, E.; Kumar, A.; Ravindran, B.; Awasthi, S.K.; Liu, T.; Duan, Y.; et al. Resource recovery and circular economy from organic solid waste using aerobic and anaerobic digestion technologies. *Bioresour. Technol.* **2020**, *301*, 122778. [CrossRef]
7. Zhang, J.; Liu, J.; Wang, Y.; Yu, D.; Sui, Q.; Wang, R.; Chen, M.; Tong, J.; Wei, Y. Profiles and drivers of antibiotic resistance genes distribution in one-stage and two-stage sludge anaerobic digestion based on microwave-H_2O_2 pretreatment. *Bioresour. Technol.* **2017**, *241*, 573–581. [CrossRef] [PubMed]
8. Wei, W.; Wu, L.; Liu, X.; Chen, Z.; Hao, Q.; Wang, D.; Liu, Y.; Peng, L.; Ni, B.-J. How does synthetic musks affect methane production from the anaerobic digestion of waste activated sludge? *Sci. Total. Environ.* **2020**, *713*, 136594. [CrossRef] [PubMed]
9. Hao, X.; Wang, X.; Liu, R.; Li, S.; Van Loosdrecht, M.C.; Jiang, H. Environmental impacts of resource recovery from wastewater treatment plants. *Water Res.* **2019**, *160*, 268–277. [CrossRef]
10. Meena, R.A.A.; Kannah, R.Y.; Sindhu, J.; Ragavi, J.; Kumar, G.; Gunasekaran, M.; Banu, J.R. Trends and resource recovery in biological wastewater treatment system. *Bioresour. Technol. Rep.* **2019**, *7*, 100235. [CrossRef]
11. Mesjasz-Lech, A. Reverse logistics of municipal solid waste—Towards zero waste cities. *Transp. Res. Procedia* **2019**, *39*, 320–332. [CrossRef]
12. Ye, Y.; Ngo, H.H.; Guo, W.; Liu, Y.; Li, J.; Liu, Y.; Zhang, X.; Jia, H. Insight into chemical phosphate recovery from municipal wastewater. *Sci. Total Environ.* **2017**, *576*, 159–171. [CrossRef] [PubMed]

13. Pradel, M.; Aissani, L. Environmental impacts of phosphorus recovery from a "product" Life Cycle Assessment perspective: Allocating burdens of wastewater treatment in the production of sludge-based phosphate fertilizers. *Sci. Total Environ.* **2019**, *656*, 55–69. [CrossRef] [PubMed]
14. Cordell, D.; Rosemarin, A.; Schröder, J.; Smit, A. Towards global phosphorus security: A systems framework for phosphorus recovery and reuse options. *Chemosphere* **2011**, *84*, 747–758. [CrossRef] [PubMed]
15. Khan, E.U.; Nordberg, Å. Membrane distillation process for concentration of nutrients and water recovery from digestate reject water. *Sep. Purif. Technol.* **2018**, *206*, 90–98. [CrossRef]
16. De Vrieze, J.; Smet, D.; Klok, J.; Colsen, J.; Angenent, L.T.; Vlaeminck, S. Thermophilic sludge digestion improves energy balance and nutrient recovery potential in full-scale municipal wastewater treatment plants. *Bioresour. Technol.* **2016**, *218*, 1237–1245. [CrossRef]
17. Ruffino, B.; Cerutti, A.; Campo, G.; Scibilia, G.; Lorenzi, E.; Zanetti, M. Thermophilic vs. mesophilic anaerobic digestion of waste activated sludge: Modelling and energy balance for its applicability at a full scale WWTP. *Renew. Energy* **2020**, *156*, 235–248. [CrossRef]
18. Lin, R.; Cheng, J.; Ding, L.; Murphy, J.D. Improved efficiency of anaerobic digestion through direct interspecies electron transfer at mesophilic and thermophilic temperature ranges. *Chem. Eng. J.* **2018**, *350*, 681–691. [CrossRef]
19. Micolucci, F.; Gottardo, M.; Valentino, F.; Cavinato, C.; Bolzonella, D. Pilot scale comparison of single and double-stage thermophilic anaerobic digestion of food waste. *J. Clean. Prod.* **2018**, *171*, 1376–1385. [CrossRef]
20. Cao, Z.; Hülsemann, B.; Wüst, D.; Illi, L.; Oechsner, H.; Kruse, A. Valorization of maize silage digestate from two-stage anaerobic digestion by hydrothermal carbonization. *Energy Convers. Manag.* **2020**, *222*, 113218. [CrossRef]
21. Srisowmeya, G.; Chakravarthy, M.; Devi, G.N. Critical considerations in two-stage anaerobic digestion of food waste—A review. *Renew. Sustain. Energy Rev.* **2020**, *119*, 109587. [CrossRef]
22. Zhang, K.; Gu, J.; Wang, X.; Yin, Y.; Zhang, X.; Zhang, R.; Tuo, X.; Zhang, L. Variations in the denitrifying microbial community and functional genes during mesophilic and thermophilic anaerobic digestion of cattle manure. *Sci. Total Environ.* **2018**, *634*, 501–508. [CrossRef] [PubMed]
23. Kim, M.; Ahn, Y.-H.; Speece, R. Comparative process stability and efficiency of anaerobic digestion; mesophilic vs. thermophilic. *Water Res.* **2002**, *36*, 4369–4385. [CrossRef]
24. Song, Y.-C.; Kwon, S.-J.; Woo, J.-H. Mesophilic and thermophilic temperature co-phase anaerobic digestion compared with single-stage mesophilic- and thermophilic digestion of sewage sludge. *Water Res.* **2004**, *38*, 1653–1662. [CrossRef]
25. Fernández-Rodríguez, J.; Pérez, M.; Romero, L.I. Semicontinuous Temperature-Phased Anaerobic Digestion (TPAD) of Organic Fraction of Municipal Solid Waste (OFMSW). Comparison with single-stage processes. *Chem. Eng. J.* **2016**, *285*, 409–416. [CrossRef]
26. Yu, Q.; Li, H.; Deng, Z.; Liao, X.; Liu, S.; Liu, J. Comparative assessment on two full-scale food waste treatment plants with different anaerobic digestion processes. *J. Clean. Prod.* **2020**, *263*, 121625. [CrossRef]
27. Sills, D.L.; Van Doren, L.G.; Beal, C.; Raynor, E. The effect of functional unit and co-product handling methods on life cycle assessment of an algal biorefinery. *Algal Res.* **2020**, *46*, 101770. [CrossRef]
28. Li, H.; Jin, C.; Zhang, Z.; O'Hara, I.M.; Mundree, S. Environmental and economic life cycle assessment of energy recovery from sewage sludge through different anaerobic digestion pathways. *Energy* **2017**, *126*, 649–657. [CrossRef]
29. Li, H.; Feng, K. Life cycle assessment of the environmental impacts and energy efficiency of an integration of sludge anaerobic digestion and pyrolysis. *J. Clean. Prod.* **2018**, *195*, 476–485. [CrossRef]
30. Zhang, B.; Su, S.; Zhu, Y.; Li, X. An LCA-based environmental impact assessment model for regulatory planning. *Environ. Impact Assess. Rev.* **2020**, *83*, 106406. [CrossRef]
31. Lanko, I.; Jenicek, P. *Anaerobic Sludge Digestion: Single System Versus Two-Stage One*; UCT Prague: Dejvice, Czech Republic, 2019; p. 1.
32. Timonen, K.; Sinkko, T.; Luostarinen, S.; Tampio, E.; Joensuu, K. LCA of anaerobic digestion: Emission allocation for energy and digestate. *J. Clean. Prod.* **2019**, *235*, 1567–1579. [CrossRef]
33. Garfi, M.; Flores, L.; Ferrer, I. Life Cycle Assessment of wastewater treatment systems for small communities: Activated sludge, constructed wetlands and high rate algal ponds. *J. Clean. Prod.* **2017**, *161*, 211–219. [CrossRef]

34. Ferrer, I.; Serrano, E.; Ponsá, S.; Vázquez, F.; Font, X. Enhancement of Thermophilic Anaerobic Sludge Digestion by 70 °C Pre-Treatment: Energy Considerations. *J. Residuals Sci. Technol.* **2009**, *6*, 11–18.
35. Riau, V.; De La Rubia, M.Á; Pérez, M. Temperature-phased anaerobic digestion (TPAD) to obtain class A biosolids: A semi-continuous study. *Bioresour. Technol.* **2010**, *101*, 2706–2712. [CrossRef] [PubMed]
36. Courtens, E.N.P.; Vlaeminck, S.E.; Vilchez-Vargas, R.; Verliefde, A.; Jauregui, R.; Pieper, D.H.; Boon, N. Trade-off between mesophilic and thermophilic denitrification: Rates vs. sludge production, settleability and stability. *Water Res.* **2014**, *63*, 234–244. [CrossRef]
37. Daelman, M.R.; Van Voorthuizen, E.M.; Van Dongen, L.G.J.M.; Volcke, E.I.P.; Van Loosdrecht, M.C.M. Methane and nitrous oxide emissions from municipal wastewater treatment—Results from a long-term study. *Water Sci. Technol.* **2013**, *67*, 2350–2355. [CrossRef]
38. Willén, A.; Rodhe, L.; Pell, M.; Jönsson, H. Nitrous oxide and methane emissions during storage of dewatered digested sewage sludge. *J. Environ. Manag.* **2016**, *184*, 560–568. [CrossRef]
39. GlobalPetrolPrices.com. Available online: https://ru.globalpetrolprices.com/energy_mix.php?countryId=188 (accessed on 9 October 2020).
40. Akgul, D.; Cella, M.A.; Eskicioglu, C. Influences of low-energy input microwave and ultrasonic pretreatments on single-stage and temperature-phased anaerobic digestion (TPAD) of municipal wastewater sludge. *Energy* **2017**, *123*, 271–282. [CrossRef]
41. Li, Y.; Han, Y.; Zhang, Y.; Fang, Y.; Li, S.; Li, G.; Luo, W. Factors affecting gaseous emissions, maturity, and energy efficiency in composting of livestock manure digestate. *Sci. Total Environ.* **2020**, *731*, 139157. [CrossRef]
42. Corominas, L.; Foley, J.; Guest, J.S.; Hospido, A.; Larsen, H.F.; Morera, S.; Shaw, A. Life cycle assessment applied to wastewater treatment: State of the art. *Water Res.* **2013**, *47*, 5480–5492. [CrossRef]
43. Massanet-Nicolau, J.; Dinsdale, R.M.; Guwy, A.; Shipley, G. Utilising biohydrogen to increase methane production, energy yields and process efficiency via two stage anaerobic digestion of grass. *Bioresour. Technol.* **2015**, *189*, 379–383. [CrossRef] [PubMed]
44. Dixon, A.; Simon, M.; Burkitt, T. Assessing the environmental impact of two options for small-scale wastewater treatment: Comparing a reedbed and an aerated biological filter using a life cycle approach. *Ecol. Eng.* **2003**, *20*, 297–308. [CrossRef]
45. Bacenetti, J.; Fiala, M. Carbon footprint of electricity from anaerobic digestion plants in Italy. *Environ. Eng. Manag. J.* **2015**, *14*, 1495–1502. [CrossRef]
46. Lizarralde, I.; Fernández-Arévalo, T.; Beltrán, S.; Ayesa, E.; Grau, P. Validation of a multi-phase plant-wide model for the description of the aeration process in a WWTP. *Water Res.* **2018**, *129*, 305–318. [CrossRef] [PubMed]
47. Uggetti, E.; Ferrer, I.; Molist, J.; García, J. Technical, economic and environmental assessment of sludge treatment wetlands. *Water Res.* **2011**, *45*, 573–582. [CrossRef]
48. Quist-Jensen, C.A.; Sørensen, M.J.; Svenstrup, A.; Scarpa, L.; Carlsen, T.S.; Jensen, H.C.; Wybrandt, L.; Christensen, M.L. Membrane crystallization for phosphorus recovery and ammonia stripping from reject water from sludge dewatering process. *Desalination* **2018**, *440*, 156–160. [CrossRef]

water

MDPI

Article

Impact of Nanoscale Magnetite and Zero Valent Iron on the Batch-Wise Anaerobic Co-Digestion of Food Waste and Waste-Activated Sludge

Ghada Kassab [1,*], Dima Khater [2], Fadwa Odeh [3], Khaldoun Shatanawi [1], Maha Halalsheh [4], Mazen Arafah [5] and Jules B. van Lier [6]

1 Civil Engineering Department, School of Engineering, The University of Jordan, 11942 Amman, Jordan; kshatanawi@ju.edu.jo
2 Department of Chemistry, Faculty of Art and Science, Applied Science Private University, 11931 Amman, Jordan; d_kahter@asu.edu.jo
3 Department of Chemistry, School of Science, The University of Jordan, 11942 Amman, Jordan; f.odeh@ju.edu.jo
4 Water, Energy and Environment Center, The University of Jordan, 11942 Amman, Jordan; halalshe@ju.edu.jo
5 Industrial Engineering Department, School of Engineering, The University of Jordan, 11942 Amman, Jordan; mazen.arafeh@ju.edu.jo
6 Section Sanitary Engineering, Department of Water Management, Faculty of Civil Engineering and Geosciences, Delft University of Technology, 2628 CN Delft, The Netherlands; J.B.vanLier@tudelft.nl
* Correspondence: Ghada.kassab@ju.edu.jo; Tel.: +962-6-535-5000

Received: 29 February 2020; Accepted: 28 April 2020; Published: 30 April 2020

Abstract: As a potential approach for enhanced energy generation from anaerobic digestion, iron-based conductive nanoparticles have been proposed to enhance the methane production yield and rate. In this study, the impact of two different types of iron nanoparticles, namely the nano-zero-valent-iron particles (NZVIs) and magnetite (Fe_3O_4) nanoparticles (NPs) was investigated, using batch test under mesophilic conditions (35 °C). Magnetite NPs have been applied in doses of 25, 50 and 80 mg/L, corresponding to 13.1, 26.2 and 41.9 mg magnetite NPs/gTS of substrate, respectively. The results reveal that supplementing anaerobic batches with magnetite NPs at a dose of 25 mg/L induces an insignificant effect on hydrolysis and methane production. However, incubation with 50 and 80 mg/L magnetite NPs have instigated comparable positive impact with hydrolysis percentages reaching approximately 95% compared to 63% attained in control batches, in addition to a 50% enhancement in methane production yield. A biodegradability percentage of 94% was achieved with magnetite NP doses of 50 and 80 mg/L, compared to only 62.7% obtained with control incubation. NZVIs were applied in doses of 20, 40 and 60 mg/L, corresponding to 10.8, 21.5 and 32.2 mg NZVIs/gTS of substrate, respectively. The results have shown that supplementing anaerobic batches with NZVIs revealed insignificant impact, most probably due to the agglomeration of NZVI particles and consequently the reduction in available surface area, making the applied doses insufficient for measurable effect.

Keywords: anaerobic co-digestion; food wastes; waste-activated sludge; nano magnetite; iron oxide nano particles; nano zero valent iron; sewage sludge; nano particles; organic wastes

1. Introduction

Anaerobic digestion (AD) converts organic matter into biogas, a renewable source of energy, and digestate, a valuable fertilizer and soil conditioner [1,2]. Due to the increasing demand on renewable energy and the progressively adopted waste management policies that request diverting wastes from landfills, the AD process has been used for the treatment of different types of organic wastes, including sewage sludge, food waste (FW), animal manure and agricultural wastes. Nevertheless, when FW is used as a single substrate, the digestion process stability can be hampered because of (i) a possible imbalance between acidogenesis and methanogenesis when high loads of rapid fermentable organic matter are applied, (ii) potential nutrients imbalance, a high carbon to nitrogen (C/N) ratio, and (iii) the high variability of FW composition [3]. A feasible and reliable approach to overcome these limitations is the use of sewage sludge as co-substrate for food waste digestion.

In the AD process, four major steps are involved, viz. hydrolysis, acidogenesis, acetogenesis and methanogenesis. Generally, the process is limited by one or two major steps, depending on the nature of the substrate. Hydrolysis is often the rate limiting step if complex organic solids are being digested. On the other hand, if the substrate is soluble organic matter, methanogenesis is generally the rate limiting step [4].

In recent years, several studies have shown that the supplementation of conductive nanoparticles has a positive effects on the anaerobic digestion process, particularly in relation to the enhancement of methane production yield and rate, the reduction in startup and recovery periods, in addition to stability improvement [5,6]. In particular, iron oxide nanoparticles (IONPs) that include magnetite, maghemite and hematite, in addition to the nano-zero-valent-iron particles (NZVIs) hold high potentials for AD enhancement and improvement of process robustness [5]. IONPs have specifically great potentials due to its high chemical stability and magnetic properties [7,8]. Most importantly, IONPs are conductive materials that may stimulate the direct interspecies electron transfer (DIET) in anaerobic digestion, in which interspecies electron transfer is not mediated by diffusive electron carriers (i.e. hydrogen or formate) but by direct transfer of electrons released from electron donating bacteria (i.e., oxidizing bacteria that can extracellularly release electrons to conductive materials) to electron capturing microorganism (i.e. methanogenic archaea) that can reduce carbon dioxide to methane using electrons transferred from the electron donating bacteria via the conductive materials [7,9]. The primary mechanism suggested to explain the enhancing behaviors of IONPs in syntrophic methanogenesis via DIET [10] is that (semi) conductive iron oxides act as electron conduits to accelerate DIET in syntrophic methanogesis. Jiang et al. [11] suggested that electron transfer takes place via the biochemical dynamic cycling among the Fe(III) (mineral)-Fe (II)-Fe (III) mineral of the (semi)conductive iron oxides. Wherein, the released electrons are accepted by Fe(III) (mineral) of iron oxides and is reduced to produce Fe(II), then the unbounded Fe(II) transfers electron to methanogens. Fe(II) itself is readsorbed and oxidized back to original structural Fe(III) (mineral) through precipitation.

Early studies tackling the impact of IONPs on anaerobic digestion have used simple substrates (such as propionate, butyrate, and methanol), thus focusing on the syntrophic methanogenesis process. Kato et al. [12] showed that supplementing rice paddy soil with (semi)conductive iron oxide NPs (magnetite, hematite), significantly stimulated methanogenesis from acetate and ethanol in terms of onset time and production rate, attributing these results to the DIET through the (semi) conductive iron oxides. Possibly, in their research, syntrophic acetate oxidation was an important methanogenic pathway, although recent research showed a direct stimulatory effect of added hydrochar to the acetoclastic methanogen *Methanosaete*, which was also ascribed to DIET [13]. Likewise, Zhang and Lu [14] showed that methane production from butyrate oxidation in lake sediments was significantly accelerated in the presence of magnetite NPs, suggesting that DIET mediated syntrophic methanogenesis. Focusing on methanogenic propionate degradation, Cruz Viggi et al. [15] showed that the supplementation of magnetite NPs to a methanogenic sludge obtained from a pilot scale anaerobic digester fed with wasted-activated sludge (WAS) resulted in a 33% enhancement in the maximum rate of methane production. Authors proposed that this stimulatory effect has most probably resulted from the establishment of a DIET with magnetite NPs serving as electron conduits between propionate oxidizing acetogens and carbon-dioxide-reducing methanogens.

The positive effects reported on the impact of conductive iron oxides on methane production yield and rate, using simple substrates, have pushed the research forward into studying the impact of IONPs on the anaerobic digestion of complex organics. Realizing that the hydrolysis of particulate organics is the rate limiting step in anaerobic digestion of complex organics, the majority of these studies have investigated the impact of IONPs on the hydrolysis and acidification processes as well as on syntrophic methanogenesis [16–18]. The outcomes of these studies have shown that magnetite NPs can positively impact the hydrolysis of complex organic materials, thus providing abundant substrates for methanogens and promoting the anaerobic digestion process. Nevertheless, the mechanisms in which such impacts are attained are still not clear yet.

In a similar manner, several studies have been previously conducted to assess the impact of NZVIs on the anaerobic digestion process. Results have shown improvement on methane production yield with the supplementation of NZVIs, attributing such enhancement to:

(i) The possibility of iron serving as an electron donor in the direct reduction of CO_2 to CH_4 by hydrogentrophic methanogens [19–21];
(ii) Shifting the fermentation pathway away from the propionic type because of the zero valent iron strong reducing property, which leads to the reduction in oxidation reduction potential (ORP) level [20,21];
(iii) NZVIs serving as a conductive material to promote DIET [21].

Additionally, hydrogen evolution from iron corrosion could enhance both hydrogentrophic methanogenesis and homoacetogenesis resulting from the increased H_2 flux as intermediary electron carrier [22–24], making the microbial consortia more susceptible for DIET. Other researchers observed that the addition of NZVIs leads to an increased conversion of complex organics to volatile fatty acids (VFAs) (i.e., improved hydrolysis and acidogenesis), which, in turn, enhanced the overall methanogenesis of complex substrates [25]. Yu et al. [26] studied the impact of NZVIs on the anaerobic digestion of WAS and found that the addition of 10 g/L NZVIs improved the hydrolysis-acidification process in which methanogenesis was completely inhibited. The results showed an 83% increase in total VFA concentration compared to the control incubation. The observed enhancement effect was accredited to enrichment of acid-forming bacteria, especially *Clostridia*. Feng et al. [22] also investigated the effect of NZVIs on the hydrolysis-acidification of waste-activated sludge when methanogenesis was inhibited. They observed an improvement in protein and polysaccharide conversion to 36.7% and 29.6%, respectively, at an NZVI dose of 4 g/L, compared to only 25.6% and 22.9% achieved in the control incubation. Moreover, the VFA production at an NZVI dose of 4 g/L was 37.3% higher compared to control incubation. Authors have attributed the enhanced hydrolysis-acidification to the increased activities of key enzymes. The results showed that the activities of protease and cellulase were increased by 85% at an NZVI dosage of 4 g/L, compared to the control incubations. The activities of acid-forming enzymes, including acetate kinase (AK), Phosphotransacetylase (PTA), butyrate kinase (BK) and phosphotransbutytrylase (PTB), were increased by 52.2% to 67.3%.

Despite the previously stated positive effects, NZVIs can cause inhibitory effects if added at elevated doses. Such inhibitory effects can be attributed to the strong reducing conditions developed at the NZVI surface, which can rapidly inactivate bacteria by causing severe damage to the cell membranes and to the respiratory activities through reductive decomposition of protein functional groups [27,28] and, possibly, to the rapid hydrogen production and accumulation that leads to the accumulation of VFAs [29].

Realizing the conceivable positive impacts of magnetite NPs and the NZVIs on the anaerobic digestion process, this research intended to study the effects of these two iron-based conductive materials on the co-digestion of food wastes and sewage sludge. This research aimed explicitly at investigating the impact of iron-based NPs on the hydrolysis process by measuring the extent of particulate organics solubilization. Moreover, the effects on the acidification and methane production yield and rate were examined as well.

2. Materials and Methods

2.1. Substrates and Inoculum

Two types of substrates were used in this study, FW and thickened WAS. FW was obtained from the main restaurant of the University of Jordan campus in Amman, Jordan; wherein, the entire quantities of kitchen wastes and dishes leftovers produced in the sampling day (approximately 60 kg) were manually assorted to eliminate non-biodegradable materials, such as aluminum cans, glasses, styrofoam and plastic products. The residual food waste that included vegetables, fruits, dairy products, starchy food, and meat-based food was subsequently mixed thoroughly and approximately a 5-kg sample was collected. To ensure homogeneity and increase specific surface area, FW samples were subsequently grinded using a kitchen grinder and stored at 4 °C for less than two days before being used in the batch tests. It is worth mentioning that FW characterization was conducted using grinded samples. Thickened WAS was obtained from the Abu-Nussier Wastewater Treatment Plant (Amman, Jordan). The treatment plant receives a yearly average flow of 3700 m^3/d of municipal wastewater with chemical oxygen demand (COD) and total suspended solids (TSS) concentrations of 960 and 470 mg/L, respectively.

Inoculum

Anaerobically digested sludge obtained from Al Shallaleh Wastewater Treatment Plant (Irbid, Jordan) was used as a source of inoculum. The anaerobic digester is a completely mixed reactor, operated at 37 °C and 20 days solids retention time. Total solids (TS) content of 21.2 g/L ± 1.4 was identified, along with volatile solids (VS) content of 15.2 g/L ± 0.85. The methanogenic activity test that was performed in triplicates using sodium acetate as the substrate at a concentration of 1 g/L and under initial substrate to inoculum ratio of 0.5gCOD/gVS [30] revealed an inoculum-specific methanogenic activity of 0.12 gCH_4-COD/gVS.d.

Before being used in the anaerobic digestion batch tests, the inoculum was pre-incubated under anaerobic conditions at 35 °C for four days to remove any residual biodegradable organic material that may have been present.

2.2. Preparation and Characteristics of the Nanoparticles

Magnetite NPs were synthesized according to the protocol described in Kang et al. [31]. A volume of 0.85 mL of 12.1 N HCl and 25 mL of purified deoxygenated water were combined and 5.2 g $FeCl_3$ along with 2.0 g $FeCl_2$ were dissolved into the solution under stirring conditions. The resulting solution was added drop wise into 250 mL of 1.5 M NaOH solution under vigorous stirring, generating an instant black precipitate of magnetite (Fe_3O_4). The precipitate was isolated using magnetic field (S-30-10 N webcraft Uster, Switzerland). Dynamic light scattering (DLS) data indicated a hydrodynamic size of 29.5 nm and polydispersity index of 0.91.

NZVI stock solution was freshly prepared by reducing ferrous chloride with sodium borohydride as reported by He et al. [32]. Briefly, 200 mL of 0.2 % w/w of sodium carboxy methyl cellulose (CMC, capping agent, Sigma –Aldrich) dissolved in deionized water was purged with high purity argon for at least 25 min. Then, 50 mL of 0.625 M of ferrous chloride tetrahydrate ($FeCl_2 \cdot 4H_2O$, 98%, BBC chemicals) was gradually added to 200 mL of 0.2% CMC under argon gas purging. Finally, 31 mL of 4 M sodium borohydride ($NaBH_4$, 98%, Sigma Aldrich) was added drop wise to the 250 mL Fe-CMC complex while the solution was vigorously shacked at 1100 rpm at room temperature. The final concentrations of NZVIs and CMC in stock solution were 0.11 M and 0.14% w/w, respectively. DLS data indicated a hydrodynamic size of 110 nm and polydispersity index of 0.85.

2.3. Anaerobic Co-Digestion Batch Tests

Anaerobic batch tests were conducted using the OxiTop® system that is designed to collect and store pressure data. The tests were performed in triplicates using batch test bottles of 1000 mL (1135 mL working volume). Necessary macro and micronutrients were added according to Angelidaki et al. [33]. The substrate that consisted of FW and WAS was added at a ratio of 1.5:1 (FW: WAS), determined based on the VS content of each type of substrate. The amounts of substrate added were calculated according to Pabon et al. [34] and based on: (i) the maximum pressure increase allowed by the OxiTop measuring system, which is 0.3 atm, (ii) a minimum substrate concentration of 1 gCOD/L, (iii) a liquid volume of 300 mL, and (iv) a maximum biomethane composition of 30%. As for the inoculum, the amounts added were based on a substrate to inoculum ratio of 1.0 $gCOD_{substrate}/gVS_{inoculum}$.

After the addition of medium solution, inoculum, substrate, and 200 mL of demineralized water, different aliquots of prepared nanoparticles stock solutions were added to reach desired nanoparticles concentrations. Afterward, demineralized water was added to reach 300 mL liquid volume and bottles were tightly sealed with OxiTop® measuring heads. Subsequently, the air in the headspace was flushed with nitrogen gas for 3 min to achieve anaerobic conditions. Then, bottles were incubated at 35 ± 1 °C with continuous shaking at 100 rpm agitation speed. It is worth mentioning that the pressure that was built up in the first two hours was released since it is mainly due to the, dissolution of gases, upon temperature increase. For bottles used as the control, only the medium solution, inoculum, substrate that includes FW and WAS and demineralized water were added.

Biogas production was measured through the detection of pressure increase at constant volume, using the OxiTop® measuring heads. The methane content in the biogas was analyzed until the test was completed; i.e., the cumulative biogas curve reached a plateau. Soluble COD and VFA concentrations were followed by taking 2 mL of liquid sample every two days. Three blank bottles, containing all additions except substrates, were used to correct for inoculum methane production.

2.4. Analytical Methods

Total and volatile solids content were analyzed according to the Standard Methods for the Examination of Water and Wastewater [35]. Total nitrogen (TN), total phosphorous (TP), total ammonia nitrogen (TAN), chloride ion (Cl^-), in addition to the pH and electrical conductivity (EC) that were measured employing a waste to distilled water ratio of 1:10, were all analyzed according to Radojevic and Bashkin [36]. Elementary analyses of carbon, oxygen, hydrogen and nitrogen were performed using an elementary analyzer (Perkin-Elmer-Vector 8910) following the manufacturer's instructions.

To determine soluble COD and VFA for FW, a room temperature water extraction was performed on 25 g of grounded FW sample in 250 mL of distilled water for 1 h under agitation. The mixture was then centrifuged (3000 rpm) for 30 min and soluble COD and VFA were determined in the supernatant after being filtered using 0.45-µm filter paper. For WAS, the samples were immediately centrifuged and soluble COD and VFA were determined in the supernatant after filtration using 0.45-µm filter paper as well. Soluble COD was determined using the HACH Lange cuvette test and evaluated by a DR3900 HACH Lange Spectrophotometer. The individual VFAs (viz. acetic, propionic and butyric acids) were analyzed using a gas chromatograph (Varian 3300) equipped with packed column (2 m length, 2 mm internal diameter) and flame ionization detector (FID). Helium was used as the carrier gas at a flowrate of 30 mL/min. The detector temperature was 250 °C. The pH of filtered sample was adjusted to less than 2 using formic acid prior to VFA analysis. Methane content in the biogas was analyzed using a gas chromatograph (PYE-NICAM 4500), equipped with packed column (1.5 m length, 4 mm internal diameter) and flame ionization detector (FID). Argon was used as a carrier gas at a flowrate of 30 mL/min. The detector temperature was 150 °C. Certified gas standards (Spantech Products) were employed for the standardization of methane. Scanning electronic microscope (SEM) images were taken using the SEM Quanta Feg 450 instrument; samples were placed on carbon stub and sputtered with gold (5 mm thickness). As for the samples' insertion, image capturing and measurement were all performed according to manufacturer instruction.

3. Calculations

3.1. Theoretical Biochemical Methane Potential

The empirical mole composition of the FW and WAS, computed from the elementary analysis, allows for determining the theoretical biochemical methane potential (BMP_{Th}) relying on the stoichiometry of the substrate anaerobic degradation reaction [37].

$$\begin{aligned} C_aH_bO_cN_d &+ \left(\tfrac{4a-b-2c-3d}{4}\right)H_2O \\ &\rightarrow \left(\tfrac{4a+b-2c-3d}{8}\right)CH_4 + \left(\tfrac{4a-b+2c+3d}{8}\right)CO_2 + dNH_3 \end{aligned} \tag{1}$$

Therefore,

$$BMP_{Th}\ (LCH_4/kgVS) = \frac{22.4 \times ((4a + b - 2c - 3d)/8) \times 1000}{12a + b + 16c + 14d} \quad (2)$$

where 22.4 correspond to the volume (L) occupied by an ideal gas under standard conditions (temperature of 273 Kelvin (K) and pressure of 101.3 kpa). The 1000 refers to the volume conversion factor from L to mL.

3.2. Theoretical COD

The theoretical COD (COD_{Th}) can by computed from the stoichiometry of the substrate oxidation reaction

$$C_aH_bO_cN_d + \left(\frac{4a + b - 2c - 3d}{4}\right)O_2 \rightarrow a\ CO_2 + \left(\frac{b - 3d}{2}\right)H_2O + dNH_3 \quad (3)$$

Therefore;

$$COD_{Th}(gCOD/gVS) = \frac{32 \times ((4a + b - 2c - 3d)/4)}{12a + b + 16c + 14d} \quad (4)$$

3.3. Experimental Biochemical Methane Potential

The experimental biochemical methane potential ($BMP_{experimental}$) was calculated based on the maximum methane production attained in batch test bottles after being corrected by the maximum methane production of the blank bottles [34].

$$BMP_{experimental}\ (LCH_4/kgVS) = \frac{\left[\left(\frac{(P_s+P_{atm})\times V}{R\times T}\right)\times CH_4\%_s\right] - \left[\left[\left(\frac{(P_{blank}+P_{atm})\times V}{R\times T}\right)\times CH_4\%_{blank}\right]\right]}{S_O} \times 22.4 \quad (5)$$

where P_s is the pressure accumulated inside the test bottle (pa), P_{atm} is the atmospheric pressure (pa), P_{blank} is the pressure accumulated in the blank bottle (pa), V is the headspace volume (m^3), T is the temperature in Kelvin (K), R is the universal gas constant, and $CH_4\%_S$ and $CH_4\%_{blank}$ are the accumulated biogas methane percent for the test and blank bottles, respectively. The S_o is the amount of substrate added in terms of VS.

3.4. Biodegradability, Hydrolysis and Acidification Percentages

Anaerobic biodegradability was assessed based on the percent of experimental BMP to the theoretical BMP.

$$Biodegradability\% = \frac{BMP_{experimental}}{BMP_{Th}} \times 100 \quad (6)$$

The hydrolysis percent was assessed based on the percent of the solubilized COD relative to the substrate initial particulate COD.

$$Hydrolysis\% = \frac{COD_{CH_4,t} + COD_{s,t} - COD_{s,t=0}}{COD_{Th,initial} - COD_{s,t=0}} \times 100 \quad (7)$$

where $COD_{CH_4,t}$ is the COD equivalent of methane produced at any time t, $COD_{s,t}$ is the soluble COD at any time t, $COD_{s,t=0}$ is the soluble COD at time t = 0 and $COD_{Th,initial}$ is the initial theoretical COD.

The acidification percent was assessed based on the percent of the acidified COD relative to the substrate initial theoretical COD.

$$Acidification\ \% = \frac{COD_{CH_4,t} + COD_{VFA,t} - COD_{VFA,t=0}}{COD_{Th,initial}} \times 100 \quad (8)$$

where $COD_{CH_4,t}$ is the COD equivalent of methane produced at any time t, $COD_{VFA,t}$ is the VFA equivalent COD at any time t, $COD_{VFA,t=0}$ is the VFA equivalent COD at time t = 0 and $COD_{Th,initial}$ is the initial theoretical COD.

3.5. Statistical Analysis

The statistical analyses were performed using the IBM SPSS statistics (version 23). Data collected for characterization of the FW and WAS were demonstrated with a mean ($\overline{x}$), standard deviation (σ) and coefficient of variation percent (CV). For the evaluation of the NZVIs and magnetite NPs' impact on the anaerobic co-digestion process, an ANOVA test with Bonferroni correction was used with a confidence interval of 95%.

3.6. Modeling of Methane Production

The modified Gompertz model was used to describe the progression of cumulative methane production [38].

$$Y(t) = Y_m \times \exp\left\{-\exp\left[\frac{\mu_m.e}{Y_m} \times (\lambda - t) + 1\right]\right\} \tag{9}$$

where Y(t) is the cumulative methane yield at a digestion time t (LCH_4/kgVS), Y_m is the maximum methane production (LCH_4/kgVS), μ_m is the maximum rate of methane production (LCH_4/ kgVS.d), λ is the lag phase time (d), and t is the incubation time (d), e = exp (1) = 2.718.

4. Results and Discussions

4.1. Characteristics of Substrates

The average data and coefficient of variations for the FW and WAS characteristics are presented in Table 1. The FW-measured pH (4.1 ± 0.5) is indeed low compared to the average values reported in the literature. Fisgativa et al. [39], who compiled and analyzed FW characteristics data from 70 studies that evaluated 120 different food wastes, revealed an FW pH value of 5.1 ± 0.7. Apparently, acidification was already instigated during storage time.

The total solid content of FW was 30%, which lies within the range stated in the literature, although it is among the highest reported [39–42]. The high VS/TS ratio (95.6%) highlights the high organic transformation potential. Nevertheless, the low level of soluble COD compared to theoretical COD (0.2) indicates the predominance of particulate COD in the FW, which can reduce the rate of degradation due to a limitation in hydrolysis.

The carbon to nitrogen ratio (C/N) of FW (17.6) is to some extent below the generally recommended level of 20–30 for an optimal anaerobic digestion process [40]. Moreover, upon co-digestion with the WAS that is generally characterized by a low C/N ratio (5.5), the resultant C/N ratio will be even lower. However, several researchers have demonstrated that the co-digestion of FW with WAS can be successfully performed under C/N ratios ranging from 8.8 to 13 [43–46].

With respect to nutrients content, the FW total Kjeldahl nitrogen (35.1 gN/kgVS) observed in this study is higher than the average values stated by Fisgativa et al. [39] but compatible to those reported by Zhang et al. [47], El Mashad and Zhang [48], Zhang et al. [49] and Agyeman and Tao [50] for types of FW similar to the one tested within this study. Phosphorous concentration (2.6 gP/kgVS) was found to be below the values reported in the literature [39,48,51], which are in the order of 5 gP/kgTS. Hence, in the context of nutrient supplementation, the comparison of the measured COD:N:P ratio (350:7.1:0.53), with what is reported in literature for successful and stable anaerobic digestion process (350:5:1) [52], confirms the deficiency of the phosphorous, for which the level obtained represents only 53% of the recommended value.

Table 1. Food waste (FW) and wasted-activated sludge (WAS) characteristics.

Parameter	n	FW		WAS	
		$\bar{x}$ (σ)	CV (%)	$\bar{x}$ (σ)	CV (%)
Physicochemical characteristics					
pH	8	4.1 (0.5)	12	7.4 (0.7)	10
TS (gTS/kg wet weight)	10	298.8 (21.6)	7		
(gTS/L)				25.2 (3.4)	13
VS (gVS/kg wet weight)	10	282.1 (24.6)	9		
(gVS/kgTS)		956.0 (10.9)	1		
(gVS/L)				21.1 (3.1)	15
Soluble COD (gO_2/kgVS) (mg/L)	8	311.2 (61.1)	20	654 (51)	8
Total Kjeldahl nitrogen (gN/kgVS)	8	35.1 (2.4)	7	107.7 (5.9)	5
Total ammonia nitrogen (gN/kg VS)	8	2.30 (0.5)	23	20 (3.0)	15
Total phosphorous (gP/kg VS)	8	2.6 (0.3)	13	20.9 (3.40)	16
Volatile fatty acids (g COD/kg VS)	8	3.7 (0.50)	12	13.10 (3.16)	24
C/N (%)		17.6		5.5	
Elementary analysis					
Carbon (%DM)	8	52.3 (4.7)	9	42.71 (3.14)	7
Hydrogen (%DM)	8	7.2 (0.8)	11	6.89 (0.6)	9
Oxygen (%DM)	8	37.1 (6.1)	16	42.5 (4.23)	10
Nitrogen (%DM)	8	3.4 (0.7)	20	7.9 (1.1)	14
Cl^- (mg/kg DM)	8	11029 (1376.8)	12	6813 (991.6)	15

In regard of the ammoniacal nitrogen, the results obtained within this study (2.30 gN/kgVS) are considerably higher than those reported in the literature [39,53]. Increased ammonia concentrations result in an increased buffering capacity for the anaerobic digestion process.

On the whole, the FW and WAS mixture obtained physicochemical characteristics, which accentuate the numerous benefits of FW co-digestion with WAS: (i) improving the moisture content for wet digestion, taking into consideration the WAS moisture content of 97.8%, (ii) enhancing the nutrients balance for bacterial growth; total Kjeldah nitrogen and total phosphorous contents in WAS equals of 107.7 gN/kgVS and 20.9 gP/kgVS, respectively, and (iii) the development of buffering capacity for the stable anaerobic digestion process.

In connection with the anaerobic biodegradability, the calculated BMP_{Th} for FW and WAS were 564.5 LCH_4/kgVS and 392.5 LCH_4/kgVS, respectively, computed based on the empirical mole composition of $C_{18.0}H_{29.4}O_{9.6}N$ for FW and $C_{6.3}H_{12.2}O_{4.7}N$ for WAS, assuming full COD conversion. The contribution of sulfur was considered negligible since the elementary analysis results, revealed below detection limit sulfur content. Also based on the empirical composition, the COD_{Th} for FW and WAS were 1.73 and 1.12 gO_2/gVS, respectively. It is worth mentioning that the COD_{Th} of the WAS deviated from the typical theoretical value of 1.42, which is linked to the overall elemental biomass composition $C_5H_7O_2N$. Apparently, the used WAS sample was more stabilized.

4.2. *Effects of Magnetite NPs and NZVIs on Hydrolysis and Acidification*

Due to the importance of hydrolysis in the kinetics of anaerobic digestion and the fact that it is usually the rate-limiting step, the impact of magnetite NPs on COD solubilization was assessed. Magnetite NP concentrations of 25, 50 and 80 mg/L were employed in anaerobic batch tests, corresponding to 13.1, 26.2 and 41.9 mg magnetite NPs/gTS of substrate calculated for the initial conditions. The results (Figure 1) show that the maximum soluble COD concentration achieved in the control incubation was 799 mg/L, which was reached after an incubation period of one day. Batches incubated with magnetite NPs had maximum soluble COD concentrations of 2280, 1852, and 1420 mg/L for magnetite NP doses of 80, 50 and 25 mg/L, respectively. Peak values were reached after six days with cumulative methane production of 49, 57 and 110 LCH_4/kgVS, for magnetite NP doses of 80, 50 and 25 mg/L, respectively. For the same incubation period (i.e., six days) the cumulative methane production in the control incubation reached 170 LCH_4/kgVS. Accordingly, to clarify whether increased soluble COD in magnetite NPs amended batches was due to the accumulation of soluble COD as a result of reduced consumption rates by methanogens (as discussed in Section 4.3 below) or due to stimulated hydrolysis, the hydrolysis percentages achieved after six days were computed. The results show that batches incubated with 80, 50 and 25 mg/L magnetite NPs, achieved hydrolysis percentages of 88%, 78%, and 55%, respectively, compared to hydrolysis percentage of 50% attained in control incubation. Hence, we concluded that magnetite NPs induced a stimulatory effect on the hydrolysis. The hydrolysis percentages achieved by the end of incubation periods in magnetite NPs amended batches were 65.1% for the 25 mg/L dose and 94.4% and 94.9% for the 50 and 80 mg/L doses, compared with 63.0% achieved in the control incubation. The positive impact induced by magnetite on hydrolysis process has been previously reported by several researchers. Zhao et al. [54], reported a twofold increase in waste-activated sludge protein hydrolysis, with magnetite (0.2 mm in diameter) dose of 10 g/L. Moreover, they have revealed an enhancement in the activity of protease and α-glycosidase enzymes by 63% and 27%, respectively. The positive impact of magnetite on hydrolysis of WAS was reported applying even bigger magnetite particle (8–12 mm), achieving a 31.2% and 11.6% increase in soluble protein and polysaccharides at a dose of 27 g/L [17]. Zhang et al. [18] reported that in batches incubated with 1 g/L magnetite NPs and with methanogenesis inhibition, the total polysaccharide decomposition was increased by 15.8% compared to the control incubation.

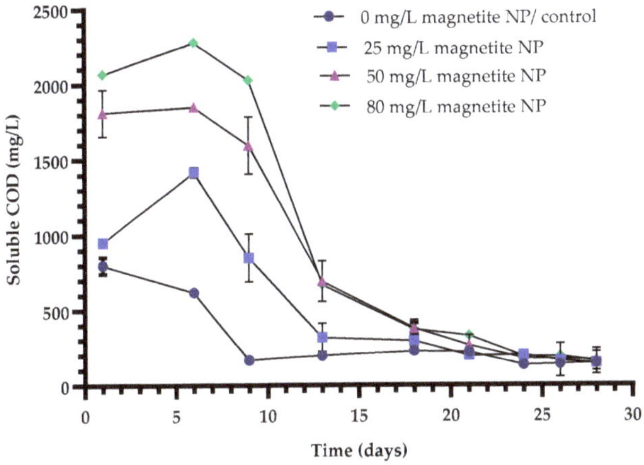

Figure 1. Effect of different magnetite NP doses on soluble COD.

Since methane yields are directly related to VFA production from substrate acidification, the impact of magnetite NPs on the availability of VFAs as precursors for methanogenesis was evaluated as well. The results show that acetate production was significantly stimulated, reaching maximum concentrations of 500, 749 and 1214 mg/L for magnetite NP doses of 25, 50 and 80 mg/L within 6 days, respectively, whereas the maximum acetate concentration in the control incubation was limited to 107 mg/L, which was reached after an incubation period of one day. Figure 2 shows that acetate was the predominant VFA, and its production is apparently directly related to the dose of the magnetite NPs. Concomitantly, methane production dropped with the increase in magnetite NPs. After the six-day incubation period, VFA concentrations started to decline, coinciding with the time at which the methane generation rate started to increase significantly, as shown in Section 4.3. To calculate the net increase in VFA production induced by magnetite NPs, the acidification percentage that takes into consideration methane production (i.e., VFA consumption) in addition to VFA generation, was computed after the six-day incubation period. The obtained results show a positive impact induced by magnetite NPs on the acidification process, with percentages reaching 54.2%, 56.6%, and 84.0% in batches incubated with magnetite NPs doses of 25, 50, and 80 mg/L, respectively. This is compared to 40.0% achieved in the control incubation. These results are compatible with those reported by Zhao et al. [54], who also reported that acetate is the main VFA generated in magnetite-amended digesters, revealing a 1.6-fold increase in acetate concentration relative to the control when amino acids were used as the substrate, and a 1.75-fold increase over the control when monosaccharides were used as the substrate. Moreover, Zhang et al. [55], who assessed acidogenesis via hydrogen yield, revealed a 1.2-fold increase in hydrogen yield compared to the control upon addition of 50 mg/L magnetite NPs.

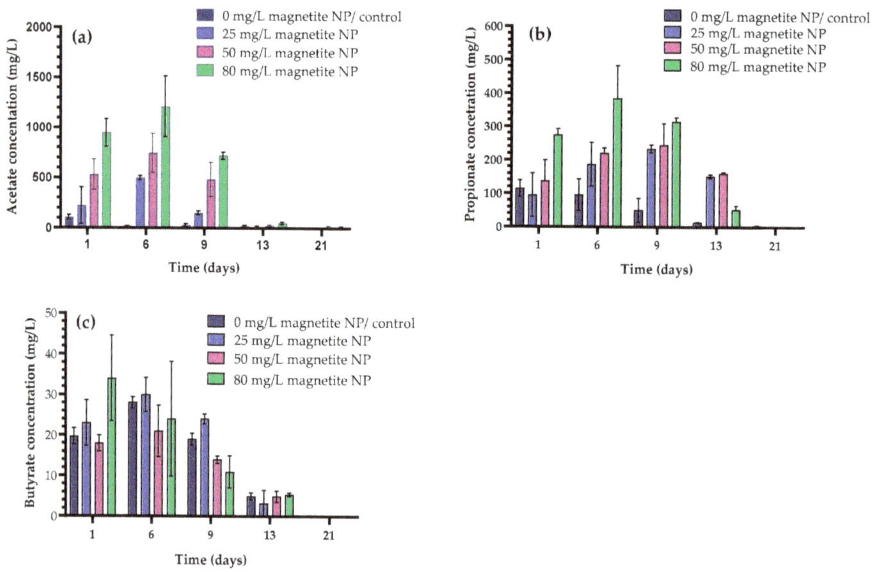

Figure 2. Effect of different magnetite NP doses on VFA production; (**a**) acetate, (**b**) propionate and (**c**) butyrate.

Magnetite NPs stimulatory impact on hydrolysis and acidification might be linked to the observed increased biomass aggregation that progressed along the incubation period, exclusively in batches incubated with magnetite NPs. Our results are congruent to the observed increased excretion of extracellular polymeric substances (EPS), brought about by magnetite supplementation, previously reported by Yin et al. [56] and Yan et al. [57]. Figure 3 shows formed aggregates after an incubation period of 8 days, along with the SEM images that were taken at the end of the incubation period. As shown by the SEM images, bacteria appeared to be aggregated and enveloped by what seems to be EPS, whereas the EPS fill the intercellular spaces within the aggregates. Observations support the hypothesis that enhanced EPS excretion may have played an essential adhesive role in the formation of aggregates and the maintenance of their integrity. Accordingly, considering the EPS sorptive capacities and their possible role in immobilizing the extracellular enzymes, the observed accelerated hydrolysis and acidification process can be explained by (i) the physical trapping of particulate and colloidal organics by means of the EPS, which leads to enhanced hydrolysis; (ii) immobilization and localization of extracellular enzymes by EPS; (iii) the minimization of hydrolysis and acidification product diffusion distances as a result of aggregation [58]. The enhanced aggregation of biomass and solid substrates implies that both enzymes and hydrolysis/acidification products remain relatively close to microbial cells, thus reducing the need for maintaining high levels of extracellular enzymes in the bulk solution and reducing the diffusive losses of products away from cells [59].

Figure 3. (**a**) Photos of the anaerobic batches, showing the biomass aggregation in batches incubated with magnetite NPs compared to the control incubation; (**1**) 80 mg/L (left) and 25 mg/L (right); (**2**) 50 mg/L, (**3**) 0 mg/L-control. (**b**) Scanning electronic microscope (SEM) images of aggregated biomass obtained from batches incubated with magnetite NPs by the end of the incubation period. (**1**) 25 mg/L; (**2**) 50 mg/L; (**3**) 80 mg/L.

Enhanced biomass aggregation resulting from magnetite nano or micro particles additions has been previously reported. Baek et al. [60,61] have studied the effect of magnetite particles (size 100–700 nm) supplementation on the anaerobic digestion of dairy effluent in a completely stirred tank reactor CSTR. Authors stated that added magnetite adhered to microbial cells' surfaces and induced microbial aggregation. Cruz Viggi et al. [15] and Li et al. [62] have studied the effect of magnetite particles on the anaerobic degradation of propionate and butyrate and showed through scanning electron micrography analysis that the magnetite particles were adsorbed on cell surfaces. This resulted in larger agglomerates, with magnetite particles appeared bridging the microbial cells. Undoubtedly, the effect of IONPs on biomass aggregation needs to be explored further so as to help clarify possible functional mechanisms of these conductive materials in enhancing aggregation.

Concerning the impact of NZVIs, results showed only slight increases in soluble COD and VFA concentrations with increased doses of NZVIs along the whole incubation period (Figures 4 and 5). Nevertheless, calculated hydrolysis and acidification percentages showed a statistically insignificant difference. Possibly, the strong clustering or agglomeration of NZVIs particles caused this negligible effect.

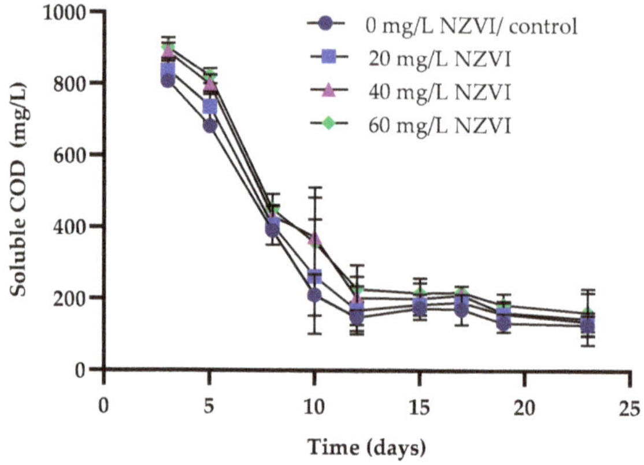

Figure 4. Effect of different nano-zero-valent-iron particle (NZVI) doses on soluble COD.

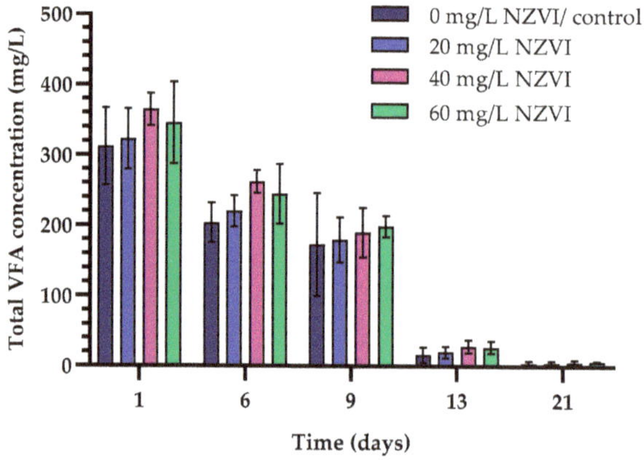

Figure 5. Effect of different nano-zero-valent-iron particle (NZVI) doses on total VFA production.

4.3. *Effects of Magnetite NPs and NZVIs on Methane Production*

The observed enhanced hydrolysis and acidification process will also impact subsequent methanogenesis. By the end of the incubation period, the cumulative methane production in the magnetite NPs amended batches (Section 4.3) reached 341.5, 478.3 and 481.5 LCH_4/kgVS, for magnetite NPs concentrations of 25, 50 and 80 mg/L, respectively. If compared with the control incubation, the batches incubated with magnetite NPs dose of 25 mg/L showed a methane production enhancement level of 7%, which was found to be statistically insignificant. With respect to the 50 and 80 mg/L magnetite NPs concentrations, results have shown a statistically significant increased methane production of 49.8% and 50.8%, respectively. These results resemble biodegradability percentages of 62.7% for the control incubation and 67.1%, 93.9% and 94.4% for incubations with 25, 50 and 80 mg/L magnetite NPs, respectively. Results undoubtedly indicated that addition of magnetite NPs increased methane production yield from anaerobic co-digestion of FW and WAS.

The addition of magnetite NPs to the batches, clearly retarded methanogenesis from the solubilized substrates (Figure 6), as also evidenced by the accumulating VFAs (Figure 2). Modeling experimental methane production data with the modified Gompertz model (Figure 6b) shows retardation periods (i.e. lag periods) of 2.8, 5.4 and 5.9 days for batches incubated with magnetite NPs doses of 25, 50, and 80 mg/L, respectively. However, after this period, the maximum methane production rate was accelerated, especially for batches incubated with magnetite NPs doses of 50 and 80 mg/L to attain an increase of 21.3% and 45.2%, relative to the control incubation (Figure 7). The initial retardation might be due to the rapid acid production resulting in a low local pH, initially inhibiting methanogenesis. Further research is required to unravel the observed phenomenon. However, if compared with the control incubation, the enhanced methane production yield in magnetite amended batches can be undoubtedly attributed to improved hydrolysis and acidification.

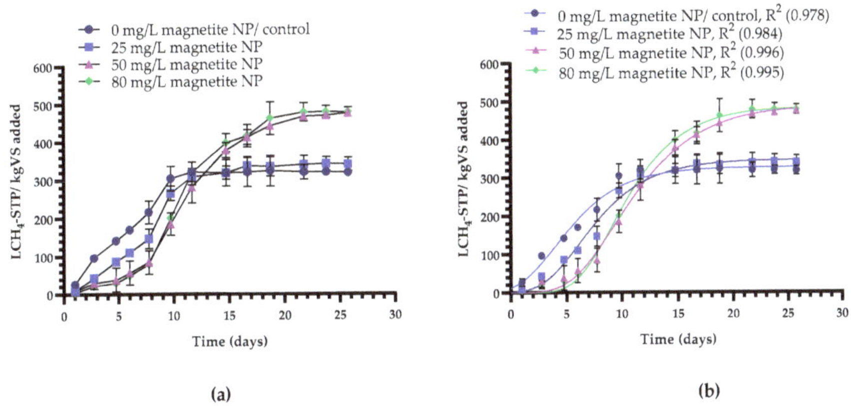

Figure 6. Cumulative methane production at different magnetite NP doses; (**a**) experimental data, (**b**) modified Gompertz model fit.

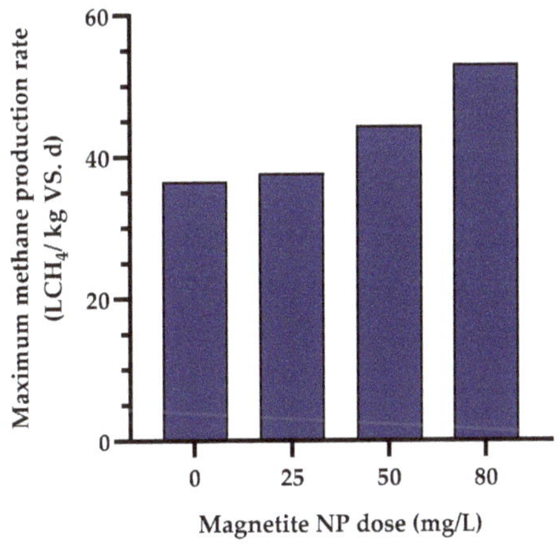

Figure 7. Effect of different magnetite NP doses on maximum methane production rates, computed from the modified Gompertz model.

Concerning the impact of NZVIs (Figure 8), the statistical analysis of computed results has revealed no measurable effect on methane generation for the three applied doses of 20, 40, and 60 mg NZVIs/L, which are equivalent to 10.8, 21.5 and 32.2 mg NZVIs/gTS of substrate, respectively. In details, the cumulative methane production at NZVIs doses of 20, 40 and 60 mg/L were 332.4, 338.3 and 343.0 LCH_4/kgVS, compared to 341.6 LCH_4/kgVS obtained with the control incubation. In literature, the impact of NZVIs on anaerobic digestion has been assessed either related to toxicity phenomena or conversion rate enhancement. The studies focusing on toxicity assessment have employed doses in the range of 55–2000 mg/L of NZVIs. Yang et al. [29] studied the impact of NZVIs on flocculent anaerobic sludge using glucose as the substrate and reported methane production inhibition levels of 20% at NZVIs doses of 1 and 10 mM (i.e. 55.9 and 558.5 mg NZVIs/L). Elevating the NZVIs dose to 30 mM (1675.5 mg NZVIs/L) resulted in 69% methane production inhibition. Authors attributed the increased inhibition to increased hydrogen accumulation resulting in reduced VFA conversion. He et al. [63] have also reported substantial methane production inhibition at NZVIs doses of 30 mM (1675.5 mg NZVIs/L) applied to flocculent sewage sludge. Jia et al. [64] have reported not only methane production inhibition but also a lag period of 15 days when treating flocculent sewage sludge with NZVIs doses of 1500 and 2000 mg/L. Studies focusing on methane production enhancement have employed NZVIs doses in the range of 1 to 10 mg NZVIs/gTS, with greater attention given to the impact on anaerobic digestion of sewage sludge. Su et al. [19] and Suanon et al. [65] have shown that the anaerobic digestion of sewage sludge in the presence of NZVIs at a concentration of 1 and 5 mg NZVIs/gTS resulted in 40.4% and 45.8% methane production yield enhancement. Substantially higher enhancement levels were achieved by Lizama et al. [21] at NZVIs doses of 3.4, 4.7 and 6.0 mg NZVIs/gTS, attaining enhancement levels of 88%, 126%, and 186%, respectively. Putting the results of this study in the context of previous studies shows that with the applied doses of the NZVIs, an enhancement of methane production is anticipated. It is considered that the insignificant impact attained may be attributed to the aggregation of NZVIs particles in form of clusters. Expectedly, aggregation of NZVIs particles will adversely affect their activity since increased size will inevitably reduce the hydrogen and ferrous iron release rates [29,66]. Consequently, the doses employed in this study may have become insufficient to lead into a notable enhancement in methane production. Actually, several previous studies have accentuated on NZVIs strong tendency for aggregation, particularly due to attractive magnetic interaction [29,67,68]. Accordingly, further investigations on the impact of NZVIs on anaerobic digestion are certainly indispensable.

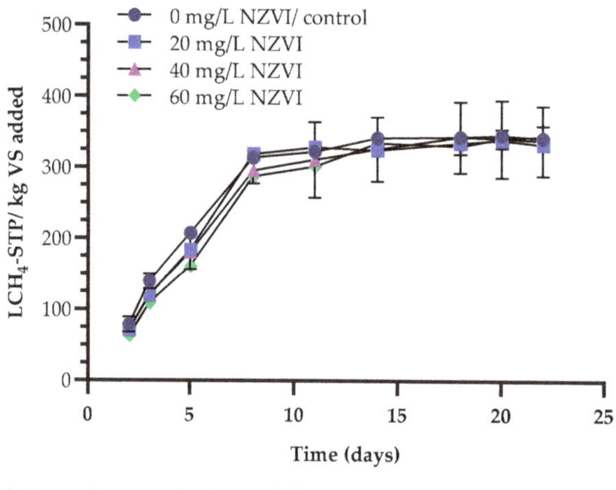

Figure 8. Cumulative methane production at different nano-zero- valent- iron particle (NZVI) doses.

5. Economic and Environmental Considerations

The obtained results, with the significant increase in methane production yield, show that supplementing the co-digestion process with magnetite NPs presents an opportunity for increased economic feasibility. On the one hand, improved methane production efficiency implies increased revenues from the elevated generation of power and heat energy. Moreover, the fact that magnetite NPs are inexpensive to produce [69,70] and can be effectively separated and reused [71] will only limitedly increase the operational costs. On the other hand, realizing effective industrial implementation necessitates a detailed economic analysis that requires further technical and scientific research to specify critical technical information, such as the maximum endurable organic loading rates, and optimum magnetite NP dose, both determined according to substrate characteristics and operating conditions.

Moreover, the results show that the addition of magnetite NPs, enhances anaerobic biodegradability percentages substantially, which consequently leads to higher volatile solids destruction. Accordingly, the quantities of generated digestate will be reduced, and thus the capital and operational cost of post digestion processes will be decreased. However, the impact of using magnetite NPs on the quality of the digestate needs to be investigated in conjunction with different operating conditions and magnetite NP doses. Particularly, if generated digestate is being considered for use as organic fertilizers. On the positive side, numerous studies have shown that IONPs have a beneficial impact on plants and lead to the improvement of crop agronomic traits [72–77]. Other studies, with the purpose of discarding toxic impacts, have shown that irrigating with water solutions containing magnetite NP concentrations as high as 1000 mg/L [78] or foliar feeding with magnetite NP solution of 10,000 mg/L [79], had no toxic impacts on plant growth. Nevertheless, and despite such promising results, the effects associated with the presence of magnetite NPs in digestate vary according to the physical and chemical characteristics of nanoparticles, soil characteristics, plant species, in addition to the rate of applications. Thus, the use of magnetite NPs on industrial scale necessitates integrated planning and management that must be supported by scientific customized studies.

6. Conclusions

This study investigated the effect of magnetite nanoparticles and nano-zero-valent-iron particles on the anaerobic co-digestion of food waste with sewage sludge. The results show that supplementing anaerobic co-digestion batches with magnetite NPs at doses of 26.2 and 41.9 mg magnetite NPs/gTS has led to a significant increase in hydrolysis percentages to a level of 94.4% and 94.8%, respectively. This is compared to 63.0% attained with the control incubation. Acidification was significantly improved as well, with acetate being the predominant VFA. Acidification percentages reached 56.6% and 84.0% in batches incubated with magnetite NP doses of 26.2 and 41.9 mg magnetite NPs/gTS, respectively, compared to only 40.0% achieved in the control incubation. The cumulative methane production yield reached 478.3 and 481.5 LCH_4/kgVS in batches incubated with 26.2 and 41.9 mg magnetite NPs/gTS, respectively. These production yields present an increase of 49.8% and 50.8% compared to the yield attained in the control incubation. Regarding the effect of nano-zero-valent-iron particles, the results show no impact, neither on methane production nor on hydrolysis or acidification.

Author Contributions: Conceptualization, G.K., F.O. and M.H.; Data curation, K.S. and M.A.; Formal analysis, J.B.v.L.; Funding acquisition, G.K.; Investigation, G.K., D.K. and F.O.; Methodology, G.K. and D.K.; Project administration, G.K.; Writing—original draft, G.K.; Writing—review & editing, M.H. and J.B.v.L. All authors have read and agreed to the published version of the manuscript.

Funding: This research was funded by Abdul Hameed Shoman Foundation/ The Scientific Research Support Fund (3/2016). Jordan.

Conflicts of Interest: The authors declare no conflict of interest

References

1. Lacovidou, E.; Ohandja, D.; Voulvoulis, N. Food waste co-digestion with sewage sludge-Realizing potential in the UK. *J. Environ. Manag.* **2012**, *112*, 267–274. [CrossRef]
2. Mehariya, S.; Patel, A.; Obulisamy, P.; Punniyakotti, E. Co-digestion of food waste and sewage sludge for methane production: Current status and perspectives. *Bioresour. Technol.* **2018**, *265*, 519–531. [CrossRef]
3. Kim, M.; Chowdhury, M.; Nakhla, G.; Keleman, M. Synergism of co-digestion of food wastes with municipal wastewater treatment biosolids. *Waste Manag.* **2017**, *61*, 473–483. [CrossRef] [PubMed]
4. Ma, J.; Frear, C.; Wang, Z.; Yu, L.; Zhao, Q.; Li, X.; Chen, S. A simple methodology for rate limiting step determination for anaerobic digestion of complex substrates and effect of microbial community ratio. *Bioresour. Technol.* **2013**, *134*, 391–395. [CrossRef] [PubMed]
5. Yang, Y.; Zhang, C.; Hu, Z. Impact of metallic and metal oxide nanoparticles on wastewater treatment and anaerobic digestion. *Environ. Sci. Process Impacts* **2013**, *15*, 39–48. [CrossRef] [PubMed]

6. Lee, Y.; Lee, D. Impact of adding metal nanoparticles on anaerobic digestion performance-A review. *Bioresour. Technol.* **2019**, *121926*, 1–9. [CrossRef]
7. Park, J.; Kang, H.; Park, K.; Park, H. Direct interspecies electron transfer via conductive materials: A perspective for anaerobic digestion applications. *Bioresour. Technol.* **2018**, *254*, 300–311. [CrossRef]
8. Sekoai, P.; Ouma, C.; du Preez, S.; Modisha, P.; Engelbrecht, N.; Bessarabov, D.; Ghimire, A. Application of nanoparticles in biofuels: An overview. *Fuel* **2019**, *237*, 380–397. [CrossRef]
9. Baek, G.; Kim, J.; Kim, J.; Lee, C. Role and potential of direct interspecies electron transfer in anaerobic digestion. *Energies* **2018**, *11*, 107. [CrossRef]
10. Xu, H.; Chang, J.; Wang, H.; Liu, Y.; Zhang, X.; Liang, P.; Huang, X. Enhancing direct interspecies electron transfer in syntrophic methanogenic association with semi conductive iron oxides: Effects and mechanisms. *Sci. Total Environ.* **2019**, *695*, 133876. [CrossRef]
11. Jiang, S.; Park, S.; Yoon, Y.; Lee, J.; Wu, W.; Phuoc Dan, N. Methanogenesis facilitated by geo-biochemical iron cycle in a novel syntrophic methanogenic microbial community. *Environ. Sci. Technol.* **2013**, *47*, 10078–10084. [CrossRef] [PubMed]
12. Kato, S.; Hashimoto, K.; Watanabe, K. Methanogenesis facilitated by electric syntrophy via (semi)conductive iron oxide minerals. *Environ. Microbiol.* **2012**, *14*, 1646–1654. [CrossRef] [PubMed]
13. Ren, S.; Usman, M.; Tsang, D.; O'thong, S.; Angelidaki, I.; Zhu, X.; Zhang, S.; Luo, G. Hydrochar-facilitated anaerobic digestion: Evidence for direct interspecies electron transfer mediated through surface oxygen-containing functional groups. *Environ. Sci. Technol.*. In press.
14. Zhang, J.; Lu, Y. Conductive Fe_3O_4 nanoparticles accelerate syntrophic methane production from butyrate oxidation in two different lake sediments. *Front. Microbiol.* **2016**, *7*, 1316–1325. [CrossRef]
15. Cruz Viggi, C.; Rossetti, S.; Fazi, S.; Paiano, P.; Majone, M. Magnetite particles triggering a faster and more robust syntrophic pathway of methanogenic propionate degradation. *Environ. Sci. Technol.* **2014**, *48*, 7536–7543. [CrossRef]
16. Zhao, Z.; Zhang, Y.; Li, Y.; Quan, X.; Zhao, Z. Comparing the mechanisms of ZVI and Fe_3O_4 for promoting waste activated sludge digestion. *Water Res.* **2018**, *1444*, 126–133. [CrossRef]
17. Pong, H.; Zhang, Y.; Tan, D.; Zhao, Z.; Zhao, H.; Quan, X. Roles of magnetite and granular activated carbon in improvement of anaerobic sludge digestion. *Bioresour. Technol.* **2018**, *249*, 666–672. [CrossRef]
18. Zhang, Y.; Zhaohui, Y.; Xu, R.; Xiang, Y.; Jia, M.; Hu, J.; Zheng, Y.; Xiong, W.; Cao, J. Enhanced mesophilic anaerobic digestion of waste sludge with the iron nanoparticles addition and kinetic analysis. *Sci. Total Environ.* **2019**, *686*, 124–133. [CrossRef]
19. Su, L.; Shi, X.; Gun, G.; Zhano, A.; Zhao, Y. Stabilization of sewage sludge in the presence of nanoscale zero valent iron (nZVI): Abatement of odor and improvement of biogas production. *J. Mater. Cycles Waste* **2013**, *15*, 461–468. [CrossRef]
20. Abdelsalam, E.; Samer, M.; Attia, Y.; Abdel Hadi, M.; Hassan, H.; Badr, Y. Influence of zero valent iron nanoparticles and magnetic iron oxide nanoparticles on biogas and methane production from anaerobic digestion of manure. *Energy* **2017**, *120*, 842–853. [CrossRef]
21. Lizama, A.; Figueiras, C.; Pedreguera, A.; Espinoza, J. Enhancing the performance and stability of the anaerobic digestion of sewage sludge by zero valent iron nanoparticles dosage. *Bioresour. Technol.* **2019**, *275*, 352–359. [CrossRef] [PubMed]
22. Feng, Y.; Zhang, Y.; Quan, X.; Chen, S. Enhanced anaerobic digestion of waste activated sludge digestion by the addition of zero valent iron. *Water Res.* **2014**, *52*, 242–250. [CrossRef] [PubMed]
23. Hao, X.; Wei, J.; van Loosdrecht, M.; Cao, D. Analyzing the mechanisms of sludge digestion enhanced by iron. *Water Res.* **2017**, *117*, 58–67. [CrossRef] [PubMed]
24. Baek, G.; Kim, J.; Lee, C. A review of the effects of iron compounds on methanogenesis in anaerobic environments. *Renew. Sust. Energy Rev.* **2019**, *113*, 109282. [CrossRef]
25. Wei, W.; Cai, Z.; Fu, J.; Xie, G.; Li, A.; Zhou, X.; Ni, B. Zero valent iron enhances methane production from primary sludge in anaerobic digestion. *Chem. Eng. J.* **2018**, *351*, 1159–1165. [CrossRef]
26. Yu, B.; Huang, X.; Zhang, D.; Lou, Z.; Yuan, H.; Zhu, N. Response of sludge fermentation liquid and microbial community to nano zero valent iron exposure in a mesophilic anaerobic digestion system. *RSC Adv.* **2016**, *6*, 24236–24243. [CrossRef]
27. Lee, C.; Kim, J.; Lee, W.; Nelson, K.; Yoon, J.; Sedlak, D. Bactericidal effect of zero valent iron nanoparticles on Escherichia coli. *Environ. Sci. Technol.* **2008**, *42*, 4927–4933. [CrossRef]

28. Kim, J.; Park, H.; Lee, C.; Kara, N.; Sedlak, D.; Yoon, J. Inactivation of *Escherichia coli* by nano particulate zero valent iron and ferrous ion. *Appl. Environ. Microbiol.* **2010**, *76*, 7668–7670. [CrossRef]
29. Yang, Y.; Guo, J.; Hu, Z. Impact of nano zero valent iron (NZVI) on methanogenic activity and population dynamics in anaerobic digestion. *Water Res.* **2013**, *47*, 6790–6800. [CrossRef]
30. Cho, Y.; Young, J.; Jordan, J.; Moon, H. Factors affecting measurement of specific methanogenic activity. *Water Sci. Technol.* **2005**, *52*, 435–440. [CrossRef]
31. Kang, Y.; Risbud, S.; Rabolt, J.; Stroeve, P. Synthesis and characterization of nanometer size Fe_3O_4 and Fe_2O_3 particles. *Chem. Mater.* **1996**, *8*, 2209–2211. [CrossRef]
32. He, F.; Zhao, D.; Liu, J.; Roberts, C. Stabilization of Fe-Pd nanoparticles with sodium carboxymethyl cellulose for enhanced transport and dechlorination of trichloroethylene in soil and groundwater. *Ind. Eng. Chem. Res.* **2007**, *46*, 29–34. [CrossRef]
33. Angelidaki, I.; Alves, M.; Bolzonella, D.; Borzacconi, L.; Campos, J.; Guwy, A.; Kalyuzhnyi, S.; Jenicke, P.; van Lier, J. Defining the biomethane potential (BMP) of solid organic wastes and energy crops: A proposed protocol for batch assays. *Water Sci. Technol.* **2009**, *59*, 927–934. [CrossRef] [PubMed]
34. Pabon, P.; Castanares, G.; van Lier, J. An oxiTop protocol for screening plant material for its biochemical methane potential (BMP). *Water Sci. Technol.* **2012**, *66*, 1416–1423. [CrossRef] [PubMed]
35. Baird, R.; Bridgewater, L. *Standard Methods for the Examination of Water and Wastewater*, 23rd ed.; American Public Health Association: Washington, DC, USA, 2017.
36. Radojevic, M.; Bashkin, V. *Practical Environmental Analysis*; RSC Publishing: Cambridge, UK, 2006.
37. Buswell, A.; Mueller, H. Mechanism of methane fermentation. *Ind. Eng. Chem.* **1952**, *44*, 550–552. [CrossRef]
38. Elbeshbishy, E.; Nakhla, G. Batch anaerobic co-digestion of proteins and carbohydrates. *Bioresour. Technol.* **2012**, *116*, 170–178. [CrossRef]
39. Fisgativa, H.; Tremier, A.; Dabert, P. Characterizing the variability of food waste quality: A need for efficient valorization through anaerobic digestion. *Waste Manag.* **2016**, *50*, 264–274. [CrossRef]
40. Zhang, C.; Su, H.; Baeyens, J.; Tan, T. Reviewing the anaerobic digestion of food waste for biogas production. *Renew. Sust. Energy Rev.* **2014**, *38*, 383–392. [CrossRef]
41. Liu, C.; Li, H.; Zhang, Y.; Liu, C. Improve biogas production from low-organic- content sludge through high-solids anaerobic co-digestion with food waste. *Bioresour. Technol.* **2016**, *219*, 252–260. [CrossRef]
42. Fisgativa, H.; Tremier, A.; Roux, S.; Bureau, C.; Dabert, P. Understanding the anaerobic biodegradability of food waste: Relationship between the typological, biochemical and microbial characteristics. *J. Environ. Manag.* **2017**, *188*, 95–107. [CrossRef]
43. Liu, L.; He, Q.; Ma, Y.; Wang, X.; Peng, X. A mesophilic anaerobic digester for treating food waste: Process stability and microbial community analysis using pyro sequencing. *Microb. Cell Factories* **2016**, *15*, 1–11. [CrossRef] [PubMed]
44. Dai, X.; Duan, N.; Dong, B.; Dai, L. High solids anaerobic co-digestion of sewage and food waste in comparison with mono digestion: Stability and performance. *Waste Manag.* **2013**, *33*, 308–316. [CrossRef] [PubMed]
45. Gou, C.; Yang, Z.; Huang, J.; Wang, H.; Xu, H.; Wang, L. Effects of temperature and organic loading rate on the performance and microbial community of anaerobic co-digestion of waste activated sludge and food waste. *Chemosphere* **2014**, *105*, 146–151. [CrossRef] [PubMed]
46. Koch, K.; Plabst, M.; Schmidt, A.; Helmreich, B.; Drewes, J. Co-digestion of food waste in a municipal wastewater treatment plant: Comparison of batch tests and full scale experiences. *Waste Manag.* **2016**, *47*, 28–33. [CrossRef]
47. Zhang, R.; El-Mashad, H.; Hartman, K.; Wang, F.; Liu, G.; Choate, C.; Gamble, P. Characterization of food waste as feedstock for anaerobic digestion. *Bioresour. Technol.* **2007**, *98*, 929–935. [CrossRef]
48. El-Mashad, H.; Zhang, R. Biogas production from co-digestion of dairy manure and food waste. *Bioresour. Technol.* **2010**, *101*, 4021–4028. [CrossRef]
49. Zhang, L.; Lee, Y.; Jahng, D. Anaerobic co-digestion of food waste and piggery wastewater: Focusing on the role of trace elements. *Bioresour. Technol.* **2011**, *102*, 5048–5059. [CrossRef]
50. Agyeman, F.; Tao, W. Anaerobic co-digestion of food waste and dairy manure: Effects of food waste particle size and organic loading rate. *J. Environ. Manag.* **2014**, *133*, 268–274. [CrossRef]
51. Zhang, Y.; Banks, C.; Heaven, S. Co-digestion of source segregated domestic food waste to improve process stability. *Bioresour. Technol.* **2012**, *114*, 168–178. [CrossRef]

52. Von Sperling, M.; Chernicharo, C. *Biological Wastewater Treatment in Warm Climate Regions*; IWA Publishing: London, UK, 2005.
53. El-Mashad, H.; Zhang, R. Co-digestion of food waste and dairy maure for biogas production. *Tasabe* **2007**, *50*, 1815–1821.
54. Zhao, Z.; Li, Y.; Yu, Q.; Zhang, Y. Ferroferric oxide triggered possible direct interspecies electron transfer between syntrophomonas and Methanosaeta to enhance waste activated sludge anaerobic digestion. *Bioresour. Technol.* **2018**, *250*, 79–85. [CrossRef] [PubMed]
55. Zhang, Z.; Guo, L.; Wang, Y.; Zhao, Y.; She, Z.; Gao, M.; Guo, Y. Application of iron oxide (Fe_3O_4) nanoparticles during the two stage anaerobic digestion with waste sludge: Impact on the biogas production and the substrate metabolism. *Renew. Energy* **2020**, *146*, 2724–2735. [CrossRef]
56. Yin, Q.; He, K.; Echigo, S.; Wu, G.; Zhan, X.; Hu, H. Ferroferric oxide significantly affected production soluble microbial products and extracellular polymeric substances in anaerobic methangenesis reactors. *Front. Microbiol.* **2018**, *9*, 2376. [CrossRef] [PubMed]
57. Yan, W.; Sun, F.; Liu, J.; Zhao, X. Enhanced anaerobic phenol degradation by conductive materials via EPS and microbial community alteration. *Chem. Eng. J.* **2018**, *352*, 1–9. [CrossRef]
58. Oh, S.; Kang, S.; Azizi, A. Electrochemical communication in anaerobic digestion. *Chem. Eng. J.* **2018**, *353*, 878–889. [CrossRef]
59. Hoffman, M.; Decho, A. Extracellular enzymes within microbial biofilms and the role of the extracellular plymer matrix. In *Microbial Extracellular Polymeric Substances: Characterization, Structure and Function*, 1st ed.; Wingender, J., Neu, T., Flemming, H., Eds.; Springer: Berlin/Heidelberg, Germany, 1999; pp. 217–227.
60. Baek, G.; Kim, J.; Lee, C. A long study on the effect of magnetite supplementation in continuous anaerobic digestion of dairy effluent-Enhancement in process performance and stability. *Bioresour. Technol.* **2016**, *222*, 344–354. [CrossRef]
61. Baek, G.; Jung, H.; Kim, J.; Lee, C. A long-term study on the effect of magnetite supplementation in continuous anaerobic digestion of dairy effluent. *Bioresour. Technol.* **2017**, *241*, 830–840. [CrossRef]
62. Li, H.; Chang, J.; Liu, P.; Fu, L.; Ding, D.; Lu, Y. Direct interspecies electron transfer accelerates syntrophic oxidation of butyrate in paddy soil enrichments. *Environ. Microbiol.* **2014**, *17*, 1533–1547. [CrossRef]
63. He, C.; He, P.; Yang, H.; Li, L.; Lin, Y.; Mu, Y. Impact of zero-valent iron nanoparticles on the activity of anaerobic granular sludge: From macroscopic to microcosmic investigation. *Water Res.* **2017**, *127*, 32–40. [CrossRef]
64. Jia, T.; Wang, Z.; Shan, H.; Liu, Y.; Lei, G. Effect of nanoscale zero-valent iron on sludge anaerobic digestion. *Resour. Conserv. Recycl.* **2017**, *127*, 190–195. [CrossRef]
65. Suanon, F.; Sun, Q.; Mama, D.; Li, J.; Dimon, B. Effect of nanoscale zero- valent iron and magnetite (Fe3O4) on the fate of metals during anaerobic digestion of sludge. *Water Res.* **2016**, *88*, 897–903. [CrossRef]
66. Gonzalez-Estrella, J.; Sierra-Alvarez, R.; Field, J. Toxicity assessment of inorganic nanoparticles to acetoclastic and hydrogentrophic methanogenic activity in anaerobic granular sludge. *J. Hazard. Mater.* **2013**, *260*, 278–285. [CrossRef] [PubMed]
67. Rosicka, D.; Sembera, J. Assessment of influence of magnetic forces on aggregation of zero valent iron nanoparticles. *Nanoscale Res. Lett.* **2011**, *6*, 10–16. [CrossRef] [PubMed]
68. Tang, S.; Lo, I. Magnetic nanoparticles: Essential factors for sustainable environmental application. *Water Res.* **2013**, *47*, 2613–2632. [CrossRef] [PubMed]
69. Xu, P.; Zeng, G.; Huang, D.; Feng, C.; Hu, S.; Zhao, M.; Lai, C.; Wei, Z.; Huang, C.; Xie, G.; et al. Use of iron oxide nanomaterials in wastewater treatment: A review. *Sci. Total Environ.* **2012**, *424*, 1–10. [CrossRef] [PubMed]
70. Wu, W.; Wu, Z.; Yu, T.; Jiang, C.; Kim, W. Recent progress on magnetic iron oxide nanoparticles: Synthesis, surface functional strategies and biomedical application. *Sci. Technol. Adv. Mater.* **2015**, *16*, 1–43. [CrossRef] [PubMed]
71. Hutchins, D.; Downey, J. Effective separation of magnetite nanoparticles within an industrial scale pipeline reactor. *Sep. Sci. Technol.* **2019**, 1–8. [CrossRef]
72. Rui, M.; Ma, C.; Hao, Y.; Guo, J.; Rui, Y.; Tang, X.; Qi, Z.; Fan, X.; Zetian, Z.; Hou, T.; et al. Iron oxide nanoparticles as a potential iron fertilizer for peanut (Arachis hypogaea). *Front. Plant Sci.* **2016**, *7*, 815. [CrossRef]

73. Li, J.; Chang, P.; Huang, J.; Wang, Y.; Yuan, H.; Ren, H. Physiological effects of magnetic iron oxide nanoparticles towards watermelon. *J. Nanosci. Nanotechnol.* **2013**, *13*, 5561–5567. [CrossRef]
74. Boutchuen, A.; Zimmerman, D.; Aich, N.; Masud, A.; Arabshahi, A.; Palchoundhury, S. Increased plant growth with haematite nanoparticles fertilizer drop and determining nanoparticles uptake in plants using multimodal approach. *J. Nanomater.* **2019**, *2019*, 6890572. [CrossRef]
75. Elfeky, S.; Mohammed, M.; Khater, M.; Osman, Y.; Elsherbini, E. Effect of magnetite nano-fertilizer on growth and yield of *Ocimum basilicum* L. *Int. J. Indig. Med. Plants* **2013**, *46*, 1286–1293.
76. Plaksenkova, I.; Jermalonoka, M.; Bankovska, L.; Gavarane, I.; Gerbreders, V.; Sledevskis, E.; Snikeris, J.; Kokina, I. Effects of Fe_3O_4 nanoparticles stress on the growth and development of rocket *Eruca sativa*. *J. Nanomater.* **2019**, *2019*, 2678247. [CrossRef]
77. Abou El-Nasr, M.; El-Hennawy, H.; El-Kereamy, A.; Abou-El-Yazied, A.; Salah Eldin, T. Effect of magnetite nanoparticles (Fe_3O_4) as nutritive supplement on pear saplings. *Middle East J. Appl. Sci.* **2015**, *5*, 777–785.
78. Govea-Alcaide, E.; Masunaga, S.; De Souza, A.; Fajardo-Rosabal, L.; Effenberger, F.; Rossi, L.; Jardim, R. Tracking iron oxide nanoparticles in plant organs using magnetic measurements. *J. Nanoparticle Res.* **2016**, *18*, 305. [CrossRef]
79. Kanjana, D. Foliar study on effect of iron oxide nanoparticles as an alternate source of iron fertilizer to cotton. *Int. J. Chem. Stud.* **2019**, *7*, 4374–4379.

water

MDPI

Article

Formulation and Characterization of a Heterotrophic Nitrification-Aerobic Denitrification Synthetic Microbial Community and its Application to Livestock Wastewater Treatment

Qi-yu Zhang [1,3], Ping Yang [2], Lai-sheng Liu [1,*] and Zeng-jin Liu [3]

1 State Key Laboratory of Simulation and Regulation of River Basin Water Cycle, China Institute of Water Resources and Hydropower Research, Beijing 100038, China; qq125535589@163.com
2 Chongqing RongJi Environmental Technology Co., Ltd., Chongqing 400000, China; liulsh@iwhr.com
3 Institute of Water Resources, North China University of Water Resources and Electric Power, Zhengzhou 450046, China; liuzengjin@ncwu.edu.cn
* Correspondence: qq125535589@126.com; Tel.: +86-138-1057-4539

Received: 4 December 2019; Accepted: 9 January 2020; Published: 13 January 2020

Abstract: There have been many studies on single strains in wastewater treatment and a new synthetic microbial community was prepared in this study, which provides a reference for the application of heterotrophic nitrification-aerobic denitrification in actual wastewater treatment. The growth period distribution of the composite bacteria was determined by plotting growth curves with different sole nitrogen sources, and the influence of the carbon source, carbon to nitrogen ratio (C/N) ratio, pH, and temperature on ammonia removal by the composite heterotrophic nitrifying-aerobic denitrifying strain was investigated. The optimal conditions for the heterotrophic nitrification process were sodium citrate as the carbon source, a C/N ratio of 10, a pH of 7, and a temperature of 30 °C, and only trace amounts of nitrate and nitrite were observed during the process. When the sequencing batch reactor (SBR) of a pig farm wastewater treatment plant was inoculated with the synthetic microbial community, the average removals of the chemical oxygen demand (COD) and ammonia nitrogen in the effluent were 92.61% and 20.56%, respectively. From the results, the synthetic microbial community was able to simultaneously perform heterotrophic nitrification-aerobic denitrification indicating great potential for full-scale applications.

Keywords: synthetic microbial community; ammonium; heterotrophic nitrification; aerobic denitrification; livestock wastewater

1. Introduction

In recent years, the livestock and poultry breeding industry has gradually moved towards specialized and large-scale centralized feeding methods [1]. Compared with traditional distributed breeding, large-scale breeding can significantly improve production efficiency, reduce production costs, and increase economic benefits. The development of large-scale farms has also led to an increase in the amount of livestock and poultry manure runoff, which has placed tremendous pressure on the ecological water environment. Livestock and poultry aquaculture wastewaters contain large amounts of suspended solids, COD, nitrogen, and phosphorus, which cause the eutrophication of water bodies and water pollution. In China, the COD and total nitrogen from the livestock wastewater account for 96% and 38% of the total agricultural wastewater, respectively [2]. In addition, the wastewater usually varies considerably in concentration and volume during different seasons or under management processes. Livestock and poultry wastewaters are not a commonly considered pollutant but contain abundant resources, including phosphorus and potassium, and the current comprehensive utilization efficiency of livestock and poultry wastewater in China is less than 60% [3,4]. By promoting the utilization of excreta and urine resources, livestock and poultry breeding wastes can become valuable products, such as biogas, organic fertilizer, and reclaimed water, which can effectively alleviate the shortage of agricultural resources in China and control non-point source pollution [5]. Efficient and economical livestock and poultry breeding industry wastewater treatment methods are urgently needed to achieve sustainable development in the modern swine breeding industry and environmental protection [6].

To remove these pollutants, biological methods are preferred in consideration of operational ease and cost [7]. A promising process called biological nitrogen removal treatment can partially and effectively dispose of both the solid and liquid fractions of manure [8]. However, there have been periodic reports on anaerobic ammonium oxidation [9] and aerobic denitrification [10], indicating that although aerobic biological processes have been applied to treat livestock and poultry breeding industry wastewaters effectively, the nitrification–denitrification process has presented challenges due to the long time required, high cost, and difficulty of management [11]. For instance, the cost of floor space and construction for the nitrification–denitrification method is high, and nitrifying bacteria grow slowly, so different conditions are required [12]. The sludge treatment process is cumbersome and easily causes secondary pollution and additional costs. Therefore, a more economical and convenient biological process is necessary [13]. The academic research on livestock and poultry breeding industry wastewaters has mainly aimed to develop a new biological nitrogen removal process and to cultivate superior strains to degrade high-concentration ammonia nitrogen wastewater. Since Robertson et al. discovered that *Thisphaera pantotropha* had heterotrophic nitrification ability in 1985, researchers have discovered a variety of heterotrophic nitrifying microorganisms with nitrification activity in soil, sludge, lake water, and the deep sea [14]. Studies have found that the heterotrophic nitrification process of *T. pantotropha* requires energy consumption. Unlike autotrophic nitrifying bacteria, *T. pantotropha* does not accumulate NO_2^--N when ammonia is oxidized under aerobic conditions [15]. With the discovery of heterotrophic nitrification-aerobic denitrification bacteria, the theory and technology of biological nitrogen removal have made important breakthroughs. Heterotrophic nitrification-aerobic denitrification has attracted extensive attention as a new type of biological nitrogen removal technology [16]. Compared with traditional biological nitrogen removal, the heterotrophic nitrification-aerobic denitrification process has higher removal efficiency of nitrogen and COD and less nitrous oxide production [17]. This process can realize the unification of nitrification and denitrification in the same time and space, greatly simplifying the process of traditional biological nitrogen removal, and therefore save operating and infrastructure costs [18]. At present, simultaneous nitrification–denitrification technology is a new method of nitrogen removal.

Removal widely occurs in the natural environment and has been successfully realized in oxidation ditches, SBR reactors and other systems [19]. Significantly, studies have shown that changes in factors such as the carbon source, dissolved oxygen, floc characteristics, and sludge concentration affect the reaction process in simultaneous nitrification and denitrification [20]. Studies have shown that many bacteria, actinomycetes, fungi, and even algae have heterotrophic nitrification capabilities. Fungi, such as *Aspergillus flavus* [21], *Penicillium* sp. [22], *Verticillium* sp. [23], *Absidia cylindrospora* [24], etc. are considered to be the most abundant and most efficient heterotrophic nitrifying microorganisms [25]; *lactobacillus*, such as *Mycobacterium*, *Nocardia*, *Micromonospora*, and algae, such as *Chlorella*, salt algae, *Phaeodactylum tricornutum* [26] also perform heterotrophic nitrification. Furthermore, many heterotrophic bacteria, such as *Pseudomonas* [27], *Alcaligenes* sp. [28], *Arthrobacter* sp. [29], and *Alcaligenes faecalis* [30] can oxidize ammonia nitrogen to nitrite nitrogen or other states [31].

As mentioned above, many single-species heterotrophic nitrification-aerobic denitrifying microorganisms and their characteristics have been discovered thus far. The synthetic microbial community can couple their efficiency, work together, and have strong environmental adaptability, and their treatment effect is better than that of single microorganisms. Nevertheless, information on the combined treatment effects and characteristics of these strains is still very limited. The purpose of this study was to develop an ammonia nitrogen degradation composite composed of heterotrophic nitrification-aerobic denitrifying strains, determine its reaction characteristics and its application in livestock wastewater treatment, and provide an experimental basis and theoretical support for future applications.

2. Materials and Methods

2.1. Media and Reagents

The beef extract peptone medium consisted of the following components: beef cream 3 g·L^{-1}, peptone 10 g·L^{-1}, and NaCl 5 g·L^{-1}. The heterotrophic nitrification medium was composed of NH_4Cl 0.38 g·L^{-1}, $C_4H_4Na_2O_4$ 5.62 g·L^{-1}, and 50 mL Vickers salt solution. The denitrification medium was made up of KNO_3 0.72 g·L^{-1}, $C_4H_4Na_2O_4$ 2.80 g·L^{-1}, KH_2PO_4 1 g·L^{-1}, $MgSO_4{\cdot}7H_2O$ 1 g·L^{-1}, and 2 mL/L trace element solution. The BTB (bromothymol blue) medium contained KNO_3 1 g·L^{-1}, L$^-$asparagine 1 g·L^{-1}, $Na_3C_6H_5O_7.5H_2O$ 8.50 g·L^{-1}, KH_2PO_4 1 g·L^{-1}, $MgSO_4{\cdot}7H_2O$ 1 g·L^{-1}, $CaCl_2$ 0.20 g·L^{-1}, $FeCl_3{\cdot}6H_2O$ 0.05 g·L^{-1}, and 5 mL/L 1% thymol blue. The LB (Luria$^-$Bertani) medium consisted of peptone 10 g·L^{-1}, yeast extract 5 g·L^{-1}, and NaCl 5 g·L^{-1}.

The Vickers salt solution contained $KH_2PO_4{\cdot}3H_2O$ 6.5 g·L^{-1}, $MgSO_4{\cdot}7H_2O$ 2.5 g·L^{-1}, NaCl 2.5 g·L^{-1}, $FeSO_4{\cdot}7H_2O$ 0.05 g·L^{-1}, and $MnSO_4{\cdot}7H_2O$ 0.04 g·L^{-1}. The trace element solution [32] contained EDTA 50 g·L^{-1}, $CaCl_2$ 5.5 g·L^{-1}, $ZnSO_4$ 2.2 g·L^{-1}, $CuSO_4{\cdot}5H_2O$ 1.57 g·L^{-1}, $FeSO_4{\cdot}7H_2O$ 5.0 g·L^{-1}, $CoCl_2{\cdot}6H_2O$ 1.61 g·L^{-1}, and $MnCl_2{\cdot}4H_2O$ 25.06 g·L^{-1}.

All the media mentioned above were adjusted to an initial pH of 7.0 to 7.2 and sterilized at 121 °C and the pressure of 0.12 MPa for 20 min.

2.2. Screening and Identification of Heterotrophic Nitrifying-Aerobic Denitrifying Strains

To obtain a high-purity single strain, it was necessary to treat pig farm sludge. Pig farm biogas slurry samples and aerobic sludge samples (10 mL) were collected, separated, efficiently transferred to a 250 mL flask containing sterilized 0.90% NaCl solution (90 mL) and glass beads and, then, shaken at 200 rpm to obtain a uniform bacterial suspension. The solution was, then, subjected to gradient dilution, and the resulting solution was inoculated on beef extract peptone medium in an incubator at 30 °C. Through five consecutive enrichment cultures, different single colonies were picked for separation and purification and then inoculated in 100 mL, which was efficiently separated into heterotrophic nitrifying liquid medium. The change in the concentration of ammonia nitrogen in the culture was qualitatively tested to complete the initial screening by observing the colour change of the Nessler reagent. Then, the obtained suspensions of different concentrations were uniformly coated on the surface of BTB medium and placed in a constant-temperature incubator to verify whether there was aerobic-denitrification activity.

The strains obtained by the primary screening were rescreened, and four strains with preferable COD degradation ability and denitrification performance were selected as the target strains and inoculated onto an inclined surface at 4 °C. The screened strains were subjected to Gram staining and observed by optical microscopy [33]. Single colonies were picked and cultured in a liquid medium to log phase, and the culture solution was used for genome extraction. The 16S rRNA sequences of the strains, amplified by universal primers 27F and 1492R, were submitted to NCBI for comparative analysis with GenBank data.

The strains selected from the biogas slurry and the aerobic sludge were prepared at a ratio of 1:1:1:1, which corresponded to the best COD degradation and denitrification performance.

2.3. Configuring the Synthetic Microbial Community and Measuring Its Growth Curve

The synthetic microbial community was transferred to NM liquid medium and cultured at 30 °C and 150 r/min for 24 h. The ammonia nitrogen removal efficiency reached an average of 91.32% at 24 h.

To determine the changes in the growth curve of the complex strain under different nitrogen sources, NH_4Cl, KNO_3, and $NaNO_2$ were selected as the only nitrogen source digestion media. During cultivation, the concentration of the synthetic microbial community, the nitrogen source concentration, and the COD concentration in the reactor were measured. The concentration of the strain was determined by regular measurements of OD_{600}, which is the absorbance of the bacterial suspension at a wavelength of 600 nm. The absorbance and the time were taken as the ordinate and the abscissa, respectively, to plot the growth curves of the synthetic microbial community, which were fitted by exponential growth curves.

2.4. *Effect of Different Factors on Heterotrophic Nitrification-Aerobic Denitrification*

To analyse the effect of different carbon sources on heterotrophic nitrification, glucose, sodium acetate, sodium succinate, and potassium sodium tartrate were selected as electron donors for the synthetic microbial community. In the experiment, the components, other than the carbon source and the nitrogen source in the NM medium, composed the basal medium. For convenience of analysis, the nitrogen source added to the medium was only ammonia nitrogen, and the carbon source to be tested was separately added to maintain a carbon to nitrogen ratio (molecular ratio) of 10.

A bevelled surface stored in a refrigerator at 4 °C was used as the source of the strains, and rings were picked into an Erlenmeyer flask containing 100 mL of NM fluid medium. The strains were cultured for 24 h under aerobic shaking at 30 °C and 150 r/min. The culture conditions were as described above.

To analyse the effect of the C/N ratio on heterotrophic nitrification, glucose was the only carbon source, and the fixed nitrogen source concentration was 156.14 $mg{\cdot}L^{-1}$ in the experiment. The carbon to nitrogen ratio (C/N) was changed by adjusting the carbon source concentration so that the C/N ratio was 1, 4, 7, 10, or 13. The strains were cultured in liquid medium at 30 °C and 150 r/min for 24 h under aerobic shaking.

To determine the optimum environmental pH or pH range of the synthetic microbial community, the pH was set to 5 different values. In this experiment, the effects of pH values of 5, 6, 7, 8, and 9 on the growth of the strains were studied.

The heterotrophic nitrification-aerobic denitrification strain was activated in culture medium until reaching log phase. The bacterial suspension was centrifuged at 8000 rpm for 10 min. The supernatant was removed and resuspended in sterile water, and this procedure was repeated 3 times. The sterilized bacterial suspension was added to a 250 mL conical flask containing 100 mL of heterotrophic nitrification medium and cultured at 20 °C, 25 °C, 30 °C, 35 °C, or 40 °C with a rotating speed of 150 rpm. After culturing in a shaking bed for 24 h, the culture solution was centrifuged to remove the cells, and the supernatant was diluted to determine the ammonia nitrogen concentration.

2.5. *Optimizing the Proportion of Strains Used to Prepare the Synthetic Microbial Community*

The synthetic microbial community was activated in the LB medium to prepare a heterotrophic nitrification-aerobic denitrification medium. According to the above experiment regarding the effects of different factors on heterotrophic nitrification-aerobic denitrification, the culture conditions were set to a carbon source of glucose, the C/N ratio of 10, pH of 7, the temperature of 30 °C, and rotation speed of 150 rpm. The ratios of the synthetic microbial community are shown in Table 1. After 48 h of culture, the supernatant taken from bacterial fluid centrifuged at 8000 rpm for 5 min was used to determine the concentration of ammonia nitrogen, and the results were compared with the results for the blank group.

Table 1. The ratio of the synthetic microbial community.

Ratio	*Pseudomonas* sp.	*Sphingobacterium* sp.	*Bacillus* sp.	*Acinetobacter* sp.
1	1	1	1	1
2	1	1	1	2
3	1	1	2	2
4	1	2	1	2
5	1	2	2	1
6	1	1	1	3
7	1	2	1	3
8	1	1	2	3
9	1	2	2	3

2.6. Application of the Synthetic Microbial Community for Pig Farm Wastewater Treatment

The wastewater came from a pig farm located in Chongqing, China, and the main pollutants contained in the manure sewage of the pig farm were organic matter, suspended matter, and ammonia nitrogen which is a high-concentration component of organic wastewater. The average flow efficiency was 2.5 m^3/d, and the COD concentration and ammonia nitrogen concentration of the influent water were 1.5 g·L^{-1} and 1 g·L^{-1}, respectively. The synthetic microbial community were added to the SBR reactor to remove the COD; the ammonia nitrogen reaction cycle was 12 h, including 9 h of influent flow and 3 h of precipitation, and the wastewater was discharged once every two cycles. The length, width, and height of the SBR reactor were 2400, 1000, and 1100 mm, respectively. Sludge loading and the sludge concentration were 0.15 kgCOD/kgMLSS·d and 4.0 kgMLSS/L, respectively. The range of the DO (dissolved oxygen) concentration threshold is relatively wide and is not clearly defined. In this experiment, the concentration of DO in the feed water was controlled on average to about 0.001 g·L^{-1} to control its stability. Experimental wastewater collection began on August 29, and COD and ammonia removal efficiency monitoring ended on October 10. The startup mode can be divided into asynchronous startup and synchronous startup according to the startup steps. The pH and temperature of the SBR reactor were approximately pH 7 to 8.2 and 30 °C, respectively. Samples were taken from the tank for the measurement of ammonia and COD in the influent and the effluent, and the data presented in this study were obtained after bioaugmentation.

2.7. Analytical Methods

pH was measured by a PHS-3B precision pH meter. Ammonium, nitrate, nitrite, and total nitrogen were measured by standard methods [34]. Ammonium was determined by the Nessler's reagent spectrophotometric method [35]. Nitrite was determined by the N-(1-naphthalene)-diaminoethane photometry method. Nitrate was measured by the ultraviolet spectrophotometric method [36], and the total nitrogen was measured by alkaline potassium persulfate digestion-UV spectrophotometry. The COD was measured by a closed reflux colorimetric method. Bacterial growth was monitored by monitoring the optical density at 600 nm (OD_{600}) using a spectrophotometer. The removal efficiency of ammonia nitrogen in the heterotrophic nitrification-aerobic denitrification process was calculated using Equation (1):

$$\eta_1 = (C_1 - C_2)/C_1 \times 100\% \quad (1)$$

where C_1 is the corresponding concentration of ammonia nitrogen at time t_1 in mg·L^{-1} and C_2 is the corresponding concentration of ammonia nitrogen at time t_2 in mg·L^{-1}.

The removal efficiency of nitrate in the heterotrophic nitrification-aerobic denitrification process was calculated using Equation (2):

$$\eta_2 = (N_1 - N_2)/N_1 \times 100\% \quad (2)$$

where N_1 is the corresponding concentration of nitrate at time t_3 in mg·L^{-1} and N_2 is the corresponding concentration of nitrate at time t_4 in mg·L^{-1}.

The removal efficiency of COD in the heterotrophic nitrification-aerobic denitrification process was calculated using Equation (3):

$$\eta_3 = (D_1 - D_2)/D_1 \times 100\% \quad (3)$$

where D_1 is the corresponding concentration of COD at time t_5 in mg·L^{-1} and D_2 is the corresponding concentration of COD at time t_6 in mg·L^{-1}. Larger values of η_1, η_2, and η_3 implied a higher capability of nitrogen and contaminant removal.

The 16S rRNA gene sequences of the strains were amplified by using the genomic DNA as the template and 16S rRNA universal primers [37]. The purified PCR products were sequenced and compared with the published data in GenBank by using BLAST [38].

3. Results and Discussion

3.1. Isolation and Identification of the Strains

After the separation, purification, and rescreening of the fifteen strains, four heterotrophic nitrifying strains were picked and named SBR-2, P6, SBR3-2, and SBR-1. The SBR-2 colony was pale yellow with a wrinkled, moist and translucent surface and irregular rectangular edges, and single colonies were short rod shaped. The P6 colony bulged with a shiny surface and milky white and opaque folded edges, and single colonies were rod shaped. The SBR3-2 colony was round and opaque and had a moist surface, and single colonies were long and rod shaped. The SBR-1 colony was yellow with a smooth and moist surface and neat edges, and single colonies were rod shaped.

The genomes of the four heterotrophic nitrifying strains were used as templates to carry out PCR amplification, and partial 16S rRNA fragments of the four strains of bacteria were obtained. A neighbour-joining tree based on the 16S rRNA gene sequences was constructed to show the phylogenetic positions of SBR3-2, SBR-1, SBR-2, P6, and representatives of some other related taxa. Bootstrap values (expressed as percentages of 1000 replications) are shown at the branch points; the scale bar represents 0.02 substitutions per nucleotide position. The measured sequences were compared to nucleic acid sequences in the GenBank database. By comparing the results and morphological characteristics mentioned above, strains SBR-2, P6, SBR3-2, and SBR-1 were identified as *Pseudomonas* sp., *Acinetobacter* sp., *Bacillus* sp., and *Sphingobacterium* sp., respectively, and their sequence similarities were all over 99%. An NJ tree was constructed using MEGA 7.0.26 (Figure 1).

The synthetic microbial community was prepared from the four strains at a ratio of 1:1:1:1. Under conditions of 30 °C and 150 r/min for 24 h, the ammonia nitrogen removal efficiency of the synthetic microbial community reached an average of 91.32%.

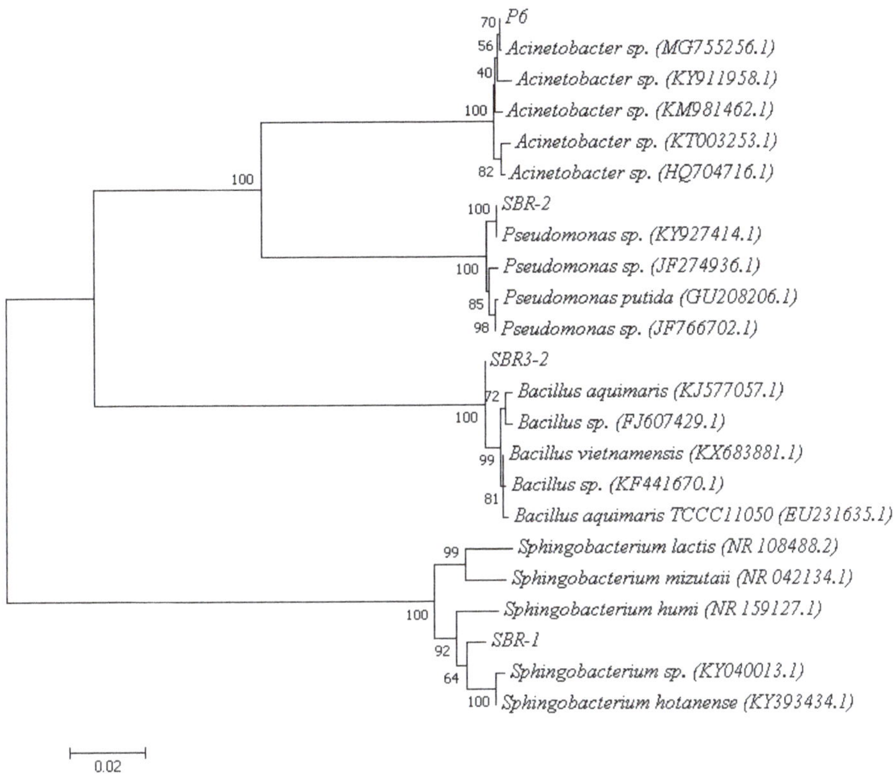

Figure 1. Phylogenetic tree based on 16S rDNA sequences of the isolates and related standard bacteria.

3.2. Growth Curves of Different Unique Nitrogen Sources

Microbial growth and reproduction mainly proceed via several stages, i.e., the adaptation, logarithmic, stable, and decay stages [39]. To understand the growth cycle of the synthetic microbial community, the growth curve of the synthetic microbial community cultured in nitrification medium with NH_4Cl as the sole nitrogen source for 24 h was calculated and is shown in Figure 2; the removal efficiencies of ammonia nitrogen and COD were 98.90% and 94.46%, respectively. From the change in absorbance at OD_{600}, the synthetic microbial community grew rapidly during the 0 to 4 h adaptation period, and the strain grew rapidly into the logarithmic growth phase after four hours After 22 h, the strain population reached a maximum and entered the stable phase. The concentration of NH_4^+ decreased slowly from 0 to 4 h, and the decrease in efficiency at 4 h was rapid. There was a very small amount of NO_2^- accumulation during the reaction, while the remaining ammonia nitrogen was partially converted into gaseous nitrogen via the desorption reaction and partially converted into other nitrogen-containing substances in solution.

The growth curve of the synthetic microbial community cultured in nitrification medium with KNO_3 as the sole nitrogen source for 24 h is shown in Figure 3, and the removal efficiencies of ammonia nitrogen and COD were 95.02% and 65.65%, respectively. The same observation method revealed that the synthetic microbial community was in the adaptation period from 0 to 2 h. After 2 h, the strains grew rapidly into the logarithmic growth phase, and the population was still increasing after 24 h. The concentration of NO_3^- decreased slowly from 0 to 2 h, and the efficiency of decline increased significantly at 2 h; NO_2^- accumulated throughout the reaction process. This result was in contrast to some heterotrophic nitrification-aerobic denitrification strains, such as LD3 [40], that have a large amount of nitrite accumulation during the reaction. Because nitrite reductase and nitrate reductase existed simultaneously in the synthetic microbial community and had high activity, the nitrite nitrogen produced during denitrification was rapidly reduced by the high-activity nitrite reductase [41].

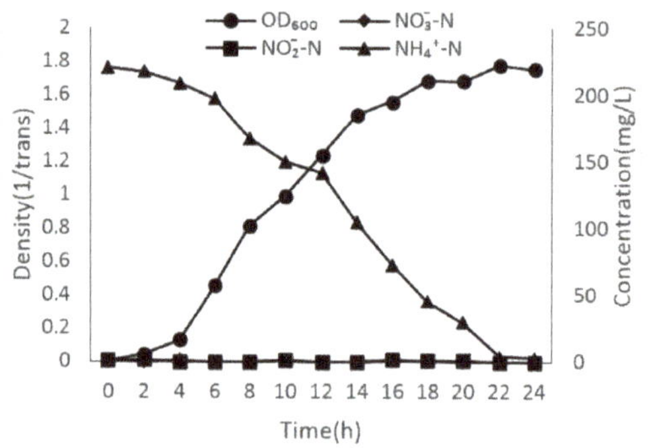

Figure 2. Growth curve of synthetic microbial community with NH_4Cl as the sole nitrogen source.

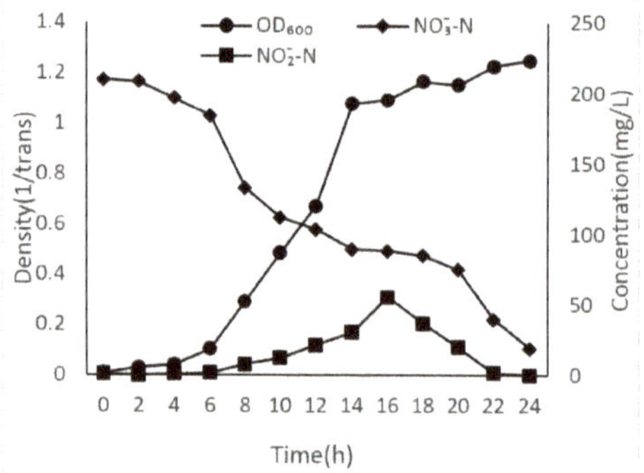

Figure 3. Growth curve of synthetic microbial community with KNO_3 as the sole nitrogen source.

The growth curve of the synthetic microbial community cultured in nitrification medium with $NaNO_2$ as the sole nitrogen source for 24 h is shown in Figure 4, and the removal efficiency of ammonia nitrogen and COD was 98.60% and 76.45%, respectively. The synthetic microbial community was in the adaptation period from 0 to 2 h. After 2 h, the strain population increased rapidly into the logarithmic growth phase, and the population reached a maximum after 22 h. Although generally, high nitrite concentrations were toxic to the strain and inhibited its growth and metabolism, some strains, such as Y-11, have a certain tolerance to high nitrite [42]. Similarly, the synthetic microbial community showed better tolerance and denitrification capacity.

The results of the batch test indicated that the synthetic microbial community had a good effect on the degradation of nitrogen and the removal of COD in media with different sole nitrogen sources. The use of NH_4Cl as the sole nitrogen source resulted in a better degradation efficiency, COD removal efficiency, and strain growth than those obtained when using $NaNO_2$/KNO_3 as the nitrogen source. The fact that nitrite had no obvious inhibitory effect on the growth and denitrification of the synthetic microbial community showed good environmental adaptability and potential application in the treatment of nitrite sewage. In summary, the synthetic microbial community can be grown for organic removal under different nitrogen sources, including ammonia nitrogen, nitrate nitrogen, and nitrite nitrogen, and the growth efficiency and organic matter removal efficiency of the synthetic microbial community are affected by the nitrogen source.

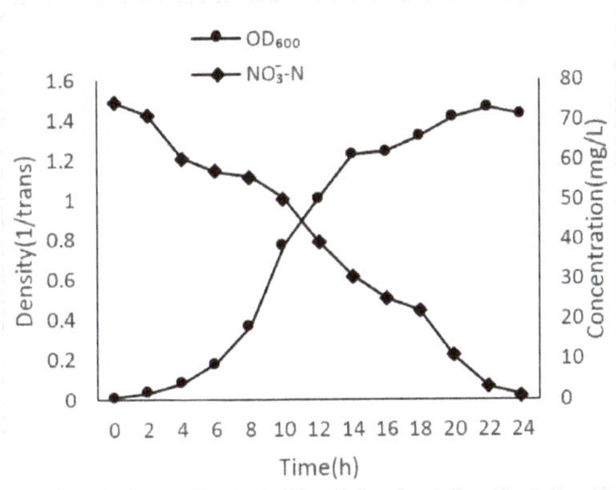

Figure 4. Growth curve of synthetic microbial community with $NaNO_2$ as the sole nitrogen source.

3.3. *Effect of Carbon Source on Heterotrophic Nitrification-Aerobic Denitrification*

The carbon source is not only an energy source for microbial nitrogen removal but also directly or indirectly affects the growth efficiency of microorganisms and the efficiency of nitrogen removal [43]. The fixed nitrogen source concentration (NH_4^+-N) was 156.14 mg·L^{-1}, and the results in Figure 5 indicate that more carbon sources could be utilized by the synthetic microbial community. In addition, potassium sodium tartrate had a very poor effect as a carbon source, and the removal efficiency of ammonia nitrogen was 98.56%, 98.75%, and 99.51%, respectively, when glucose, sodium acetate, and sodium succinate were used as the carbon sources. The effect of sodium tartrate as the sole carbon source is poor because the concentration of different carbon sources in our experiments remains the same. The reaction process using sodium tartrate as a carbon source generally requires a higher carbon source concentration, which reflects the effect deviation. In summary, different carbon sources affect the heterotrophic nitrification ability of the composite bacteria. A suitable carbon source is one of the keys to improving the ammonia nitrogen removal efficiency. The most suitable substance for heterotrophic nitrification by the synthetic microbial community was sodium succinate in this experiment.

3.4. *Effect of the C/N Ratio on Heterotrophic Nitrification-Aerobic Denitrification*

The NH_4^+-N removal percentages were significantly different among C/N ratios of one to 13, as shown in Figure 5. According to the experimental data, the ammonia nitrogen removal efficiency was only 15.92% when the C/N ratio was one, and the main reason was that an insufficient carbon supply could damage both microbial growth and the denitrification of electron donors [44]. As the C/N ratio increased, the ammonia nitrogen removal efficiency increased gradually until the C/N ratio reached 10, and the ammonia nitrogen removal efficiency reached a maximum at 98.98%. However, when the C/N ratio increased from 10 to 13, the ammonia nitrogen removal efficiency decreased, and the explanation was as follows: the higher carbon-nitrogen ratio made the carbon source content so high that some organic matter was directly embedded into the enzyme structure, affecting enzyme activity [45]. Experiments have shown that the amount of organic carbon plays an essential role in cell growth and denitrification. If the carbon source was insufficient, there was not enough electron flow to provide enough energy for the growth of the cells, and thus the denitrification capacity was relatively low. If the carbon source provided had a much higher content than the demand of the cells, the carbon source was no longer a limiting factor; the growth and metabolic activity of the cells were in a stable phase and can even have undergone reverse growth, and the denitrification capacity was stabilized or decreased. The experimental results showed that the optimal carbon to nitrogen ratio was 10, and the findings for other denitrifying bacteria, such as Vibrio diabolic SF16, seem consistent [46].

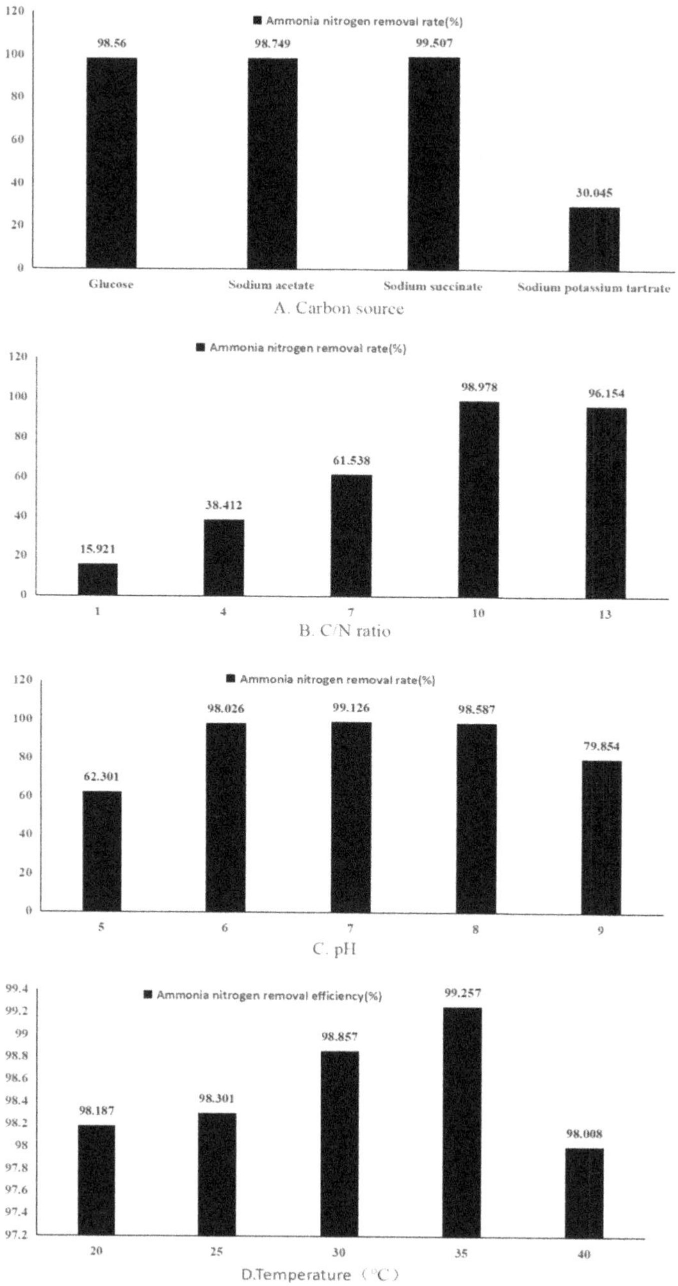

Figure 5. Effect of factors on the growth and nitrification ability of the synthetic microbial community. Carbon source (**A**), ratio of carbon to nitrogen (C/N) (**B**), pH (**C**), and temperature (**D**).

3.5. Effect of Changes in pH on Heterotrophic Nitrification-Aerobic Denitrification

The pH in the environment has a significant influence on the life activities of microorganisms. The primary role of pH is to cause a change in the charge of the cell membrane, thereby affecting the absorption of nutrients by the microorganism and affecting the enzyme activity [47]. As shown in Figure 5, the synthetic microbial community demonstrated acid and alkali inhibition at pH five and pH nine, respectively. The strain achieved the highest ammonia removal efficiency of 99.13% at pH seven, and the ammonia nitrogen removal efficiency of the synthetic microbial community could reach 98% or more in the range of pH six to nine.

In general, most heterotrophic nitrification-aerobic denitrification bacteria prefer a neutral or slightly alkaline environment, and the optimal pH range is six to nine [48]. However, when the culture medium is different, the optimum pH of heterotrophic nitrifying bacteria will also change. In beef extract peptone medium, the nitrification activity of the strain was not very sensitive, but in glucose-ammonium acetate medium, the nitrification activity was affected by changes in pH.

3.6. Effect of Changes in Temperature on Heterotrophic Nitrification-Aerobic Denitrification

The ammonia nitrogen removal efficiency of the synthetic microbial community at different temperatures is shown in Figure 5. In the range 20–40 °C, the denitrification performance of the synthetic microbial community was very good, and a value of 98% was maintained. Although the synthetic microbial community maintained an ammonia nitrogen removal efficiency of 98.01%, it had a slight decline at 40 °C. What caused this decline was that under high-temperature conditions, the active substances, such as enzymes, in microorganisms are denatured [49], and some cell functions are decreased or even terminated. These results suggested that average temperatures did not play an essential role in the process of heterotrophic nitrification of the synthetic microbial community. Taking cost-effectiveness into consideration, 30 °C was the most suitable reaction temperature.

3.7. Optimizing the Proportions of the Strains Used to Prepare the Synthetic Microbial Community

It can be concluded from Table 2, that the ammonia nitrogen removal efficiency and nitrate removal efficiency dropped significantly upon increasing the proportion of *Acinetobacter* sp., even to the relative contents corresponding to ratios two, six, and seven. When the ratio of *Pseudomonas* sp. and *Bacillus* sp. was kept constant, while increasing the proportion of *Sphingobacterium* sp. and *Acinetobacter* sp., the ammonia nitrogen removal rate and nitrate removal efficiency increased, and the maximum value was obtained at ratio number four. This result could have been obtained because the metabolites of a particular strain could have stimulated the growth of other strains or affect their functions, and thus affected the function of the whole system [50]. Comparing experimental group one and experimental group two,, experimental group eight and experimental group nine, a single increase of the ratio of *Acinetobacter* sp. or a single decrease of the ratio of *Pseudomonas* sp. is harmful to the entire system. However, from the perspective of the interaction between microorganisms, it is suitable for the whole system to reduce the ratio of *Pseudomonas* sp. and *Bacillus* sp. and increase the ratio of *Acinetobacter* sp. and *Sphingobacterium* sp. simultaneously. In conclusion, the synthetic microbial community ratio with the highest ammonia nitrogen removal efficiency is *Pseudomonas* sp.: *Sphingobacterium* sp.: *Bacillus* sp.: *Acinetobacter* sp. = 1:2:1:2 (volume ratio).

Table 2. The effect of the composite ratio of synthetic microbial community on removal efficiency of ammonia nitrogen and nitrate.

Ratio	Ammonia Nitrogen Removal Efficiency (%)	Nitrate Emoval Efficiency (%)
1	91.32	98.05
2	80.77	90.75
3	84.33	94.35
4	96.75	99.30
5	96.75	86.10
6	82.48	85.50
7	77.20	85.65
8	76.91	97.95
9	70.49	85.65

3.8. Application of the Synthetic Microbial Community for Pig Farm Wastewater Treatment

In this application, a bioenhancement scheme was mainly adopted, and the optimized synthetic microbial community was added to the SBR reactor. To enhance the ammonia nitrogen and nitrate removal performance, the reaction conditions were adjusted to the optimal conditions obtained by experimental analysis, i.e., glucose as a carbon source, a C/N ratio of 10, a pH of 7, and temperature of 30 °C. The SBR reactor had been operating stably for 40 days, and the average ammonia nitrogen concentration of the effluent water, the average ammonia nitrogen removal efficiency, and the average COD removal efficiency were 42.53 mg·L^{-1}, 92.61%, and 76.45%, respectively, as shown in Figures 6 and 7. To confirm whether ammonia nitrogen eventually became nitrogen gas, the total nitrogen was detected in the effluent. The total nitrogen value per day was slightly higher than that of ammonia nitrogen, which proved that the ammonia-degrading composite agent had indeed played its intended role and completed the biological denitrification process of ammonia from nitrogen to nitrogen. It is worthwhile mentioning that due to the rain, the effluent was diluted, and the ammonia nitrogen in the inlet water was lower than usual. According to the analysis, the excessively high ammonia nitrogen concentration in the inlet water would not be suitable for microbial reactions. After the ammonia nitrogen concentration was appropriately reduced, the removal rate was significantly increased. These results indicate that the use of the synthetic microbial community would result in a significant improvement in the biological nitrogen removal of pig farm wastewater.

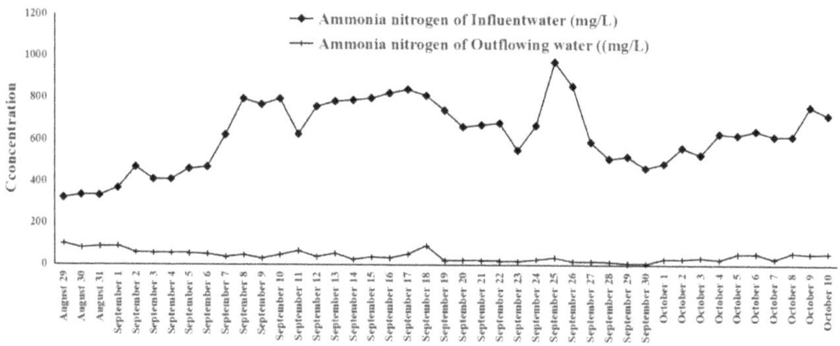

Figure 6. Ammonia nitrogen concentration change in the SBR reactor.

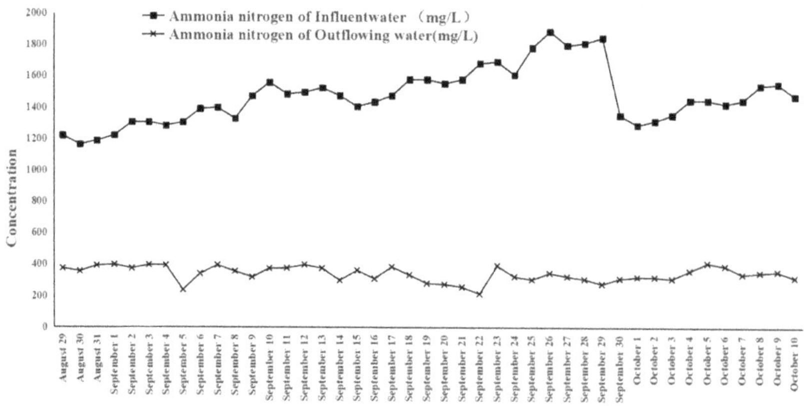

Figure 7. COD concentration change in the SBR reactor.

4. Conclusions

In this study, four heterotrophic nitrification aerobic denitrification strains, SBR2, P6, SBR3-2, and SBR-1, were isolated and identified as *Pseudomonas* sp., *Acinetobacter* sp., *Bacillus* sp., and *Sphingobacterium* sp., respectively. The synthetic microbial community was composed of the four above strains in a ratio of 1:1:1:1, and its growth curve was drawn under different nitrogen sources. By adjusting the reaction conditions, including the carbon source, carbon to nitrogen ratio, pH, and temperature, and considering the economic benefits and nitrogen removal performance, the reaction conditions of the synthetic microbial community were determined to be sodium succinate as a carbon source at C/N of 10, pH of 7, and 30 °C. The configuration scheme was optimized by adjusting the ratio of the four strains to 1:2:1:2 and detecting their denitrification performance, and the final synthetic microbial community was obtained and applied in pig farm wastewater treatment. In summary, the synthetic microbial community is a promising candidate for extensive application in various pollution control systems, including livestock wastewater and the aquaculture industry, and the next step is to determine the growth characteristics and denitrification performance of the synthetic microbial community under extreme temperature, pH, and ammonia nitrogen concentration values.

Author Contributions: L.-s.L. contributed to the conception of the study; P.Y. contributed significantly to analysis and manuscript preparation; Q.-y.Z. performed the data analyses and wrote the manuscript; Z.-j.L. helped perform the analysis with constructive discussions. All authors have read and agreed to the published version of the manuscript.

Funding: This research was funded by Ministry of Science and Technology of China grant number No. 2018YFC0408103.

Acknowledgments: This research was supported by a grant from the National Key Research and Development Program of China on Domestic Sewage Treatment Technology Development (no. 2018YFC0408103).

Conflicts of Interest: The authors declare no conflict of interest.

References

1. Peng, L.; Bai, Y. Numerical study of regional environmental carrying capacity for livestock and poultry farming based on planting-breeding balance. *J. Environ. Sci.* **2013**, *25*, 1882–1889. [CrossRef]
2. Zheng, T.; Li, P.; Ma, X.; Sun, X.; Wu, C.; Wang, Q.; Gao, M. Pilot-scale multi-level biological contact oxidation system on the treatment of high concentration poultry manure wastewater. *Process Saf. Environ. Prot.* **2018**, *120*, 187–194. [CrossRef]
3. Raha, D.; Mahanta, P.; Clarke, M.L. The implementation of decentralised biogas plants in Assam, NE India: The impact and effectiveness of the national biogas and manure management programme. *Energy Policy* **2014**, *68*, 80–91. [CrossRef]
4. Chen, R.; Gao, H.Y.; Bo, X.Q.; Shi, R.G.; Zhang, Y.G.; Ma, B.L. Study and application of treatment technology on wastewater from livestock farm. *J. Agro-Environ. Sci.* **2006**, *25*, 374–377.
5. Li, Y.; Xu, Z.; Xie, M.; Zhang, B.; Li, G.; Luo, W. Resource recovery from digested manure centrate: Comparison between conventional and aquaporin thin-film composite forward osmosis membranes. *J. Membr. Sci.* **2019**, *593*, 1136–1145. [CrossRef]
6. Weia, X.M.; Lina, C.; Duan, N.; Peng, Y.X.; Ye, Z.Y. Application of aerobic biological filter for treating swine farms wastewater. *Procedia Environ. Sci.* **2010**, *2*, 1569–1584. [CrossRef]
7. Xu, A.; Yao, J.; Yu, L.; Lv, S.; Wang, J.; Yan, B.; Yu, Z. Mutation of *Gluconobacter* oxydans and *Bacillus* megaterium in a two-step process of l-ascorbic acid manufacture by ion beam. *Appl. Microbiol.* **2004**, *96*, 1317–1323. [CrossRef]
8. Hellinga, C.; Schellen, A.A.; Mulder, J.W.; van Loosdrecht, M.C.M.; Heijnen, J.J. The sharon process: An innovative method for nitrogen removal from ammonium-rich wastewater. *Water Sci. Technol.* **1998**, *37*, 135–142. [CrossRef]
9. Fux, C.; Boehler, M.; Huber, P.; Brunner, I.; Siegrist, H. Biological treatment of ammonium-rich wastewater by partial nitrification and subsequent anaerobic ammonium oxidation (anammox) in a pilot plant. *J. Biotechnol.* **2002**, *99*, 295–306. [CrossRef]

10. Meiberg, J.B.; Bruinenberg, M.P.M.; Harder, W. Effect of dissolved oxygen tension on the metabolism of methylated amines in *Hyphomicrobium* X in the absence and presence of nitrate: Evidence for aerobic denitrification. *J. Gen. Microbiol.* **1980**, *120*, 453–463. [CrossRef]
11. Cao, L.P.; Zhou, T.; Li, Z.H.; Wang, J.J.; Tang, J.; Ruan, R.; Liu, Y.H. Effect of combining adsorption-stripping treatment with acidification on the growth of *Chlorella vulgaris* and nutrient removal from swine wastewater. *Bioresour. Technol.* **2018**, *263*, 10–16. [CrossRef] [PubMed]
12. Liu, C.; Hou, L.; Liu, M.; Zheng, Y.; Yin, G.; Han, P.; Dong, H.; Gao, J.; Gao, D.; Chang, Y.; et al. Coupling of denitrification and anaerobic ammonium oxidation with nitrification in sediments of the Yangtze Estuary: Importance and controlling factors. *Estuar. Coast. Shelf Sci.* **2019**, *220*, 64–72. [CrossRef]
13. Gupta, A.B.; Gupta, S.K. Simultaneous carbon and nitrogen removal from high strength domestic wastewater in an aerobic RBC biofilm. *Water Res.* **2001**, *35*, 1714–1722. [CrossRef]
14. Spiller, H.; Dietsh, E.; Kessler, E. Intracellular appearance of nitrite and nitrate in nitrogen-starved cells of *Ankistrodesmus braunii*. *Planta* **1976**, *129*, 175–181. [CrossRef] [PubMed]
15. Robertson, L.A.; Van Niel, E.W.; Torremans, R.A.; Kuenen, J.G. Simultaneous nitrification and denitrification in aerobic chemostat cultures of *Thisphaera pantotropha*. *Appl. Environ. Microbiol.* **1988**, *54*, 2812–2813. [CrossRef]
16. Chen, Q.; Ni, J. Heterotrophic nitrification–aerobic denitrification by novel isolated bacteria. *J. Ind. Microbiol. Biotechnol.* **2011**, *38*, 1305–1310. [CrossRef]
17. Yin, M.R.; Wang, P.; Liu, J.N.; Wang, L.; Li, A.B. Screening and identification of a heterotrophic nitrification-aerobic denitrification strain with N_2O emission control ability. *Res. Environ. Sci.* **2010**, *23*, 515–520.
18. Okamoto, K.; Washiyama, K.; Harada, Y. Renovation of an extended aeration plant for simultaneous biology removal of nitrogen and phosphorous using oxic-anaerobic-oxic process. *Water Sci. Technol.* **1990**, *22*, 61–68.
19. Kim, J.K.; Park, K.J.; Cho, K.S.; Nam, S.W.; Park, T.J.; Bajpai, R. Aerobic nitrification-denitrification by heterotrophic *Bacillus* strains. *Bioresour. Technol.* **2005**, *96*, 1897–1906. [CrossRef]
20. Gupta, A.B. *Thiosphaeern pantotropha-sulphur* bacterium capable for simultaneous hetertrophic nitrification and aerobic denitrificatio. *Enzyme Microb. Technol.* **1997**, *21*, 589–595. [CrossRef]
21. White, J.P.; Johnson, G.T. Aflatoxin production correlated with nitrification in *Aspergillus flavus* group species. *Mycologia* **1982**, *74*, 718–723. [CrossRef]
22. Killham, K.; Prosser, J.I. *Heterotrophic Nitrification*; Nitrification IRL Press: Oxford, UK, 1986; Volume 20, pp. 117–126.
23. Lang, E.; Jaqnow, G. Fungi of a forest soil nitrifying at low pH values. *FEMS. Microbiol. Ecol.* **1986**, *38*, 257–265. [CrossRef]
24. Stroo, H.F.; Klein, T.M.; Alexander, M. Heterotrophic nitrification in an acid forest soil and by an acid-tolerant fungus. *Appl. Environ. Microbiol.* **1986**, *52*, 1107–1111. [CrossRef]
25. Pedersen, H.; Dunkin, K.A.; Firestone, M.K. The relative importance of autotrophic and heterotrophic nitrification in a conifer forest soil as measured by 15N tracer and pool dilution techniques. *Biogeochemistry* **1999**, *44*, 135–150. [CrossRef]
26. Yu, Y.; Liu, Y.; Han, Y.Y.; Ji, P.; Yang, H.B. The Primary Studies on Ammonia-nitrogen Removal from Fisheries Process Wasterwater by *Chlorella vulgaris*. *Biotechnology* **2006**, *16*, 73–74.
27. Castignetti, D.; Hollocher, T. Heterotrophic nitrification among denitrifiers. *Appl. Environ. Microbiol.* **1894**, *47*, 620–623. [CrossRef]
28. Castignetti, D.; Yanong, R.; Ramzinski, R.G. Substrate diversity of an active heterotrophic nitrifier, an *Alcaligenes* sp. *Can. J. Microbiol.* **1985**, *31*, 441–445. [CrossRef]
29. Witze, L.K.; Overbeck, H. Heterotrophic nitrification by *Arthrobacter* sp. (strain 9006) as influenced by different cultural conditions, growth state and acetate metabolism. *Arch. Microbiol.* **1979**, *122*, 137–143. [CrossRef]
30. Anderson, I.C.; Poth, M.; Homstead, J.; Burdige, D. A comparison of NO and N_2O production by the autotrophic nitrifier Nitrosomonas europaea and the heterotrophic nitrifier *Alcaligenes faecalis*. *Appl. J. Environ. Microbiol.* **1993**, *59*, 3525–3533. [CrossRef]
31. Kuenen, J.G.; Robertson, L.A. Combined nitrification-dentrification processes. *FEMS Microbiol. Rev.* **1994**, *15*, 109–117. [CrossRef]
32. Li, P.; Zheng, Y.L.; Chen, S.L. Identification of an aerobic denitrifying bacterium and its potential application in wastewater treatment. *Chin. J. Appl. Environ. Biol.* **2005**, *11*, 600–603.

33. Xie, J.; Dong, G.; Liu, Z. Microbial sample preparation method by scanning electron microscope. *J. Chin. Electron Microsc. Soc.* **2005**, *24*, 440.
34. APHA. *Standard Methods for the Examination of Water and Wastewater*, 19th ed.; American Public Health Association: Washington, DC, USA, 1995.
35. Zhang, Q. Research on key issues in determination of ammonia nitrogen in water and wastewater by Nessler's reagent spectrophotometry. *Environ. Eng.* **2009**, *27*, 85–88.
36. Mahmood, Q.; Zheng, P.; Hayat, Y.; Jin, R.C.; Azim, M.R.; Jilani, G.; Islam, E.; Ahmed, M. Effect of nitrite to sulfide ratios on the performance of anoxic sulfide oxidizing reactor. *Arab. J. Sci. Eng.* **2009**, *34*, 45–54.
37. Zhu, Y.Q.; Wei, J.B. Isolation and characterization of an aerobic denitrifier. *Microbiology* **2009**, *36*, 616–619.
38. Madueno, L.; Coppotelli, B.M.; Alvarez, H.M.; Morelli, I.S. Isolation and characterization of indigenous soil bacteria for bioaugmentation of PAH contaminated soil of semiarid Patagonia. *Argent. Int. Biodeterior. Biodegrad.* **2011**, *65*, 345–351. [CrossRef]
39. Yates, G.T.; Smotzer, T. On the lag phase and initial decline of microbial growth curves. *J. Theor. Biol.* **2007**, *244*, 511–517. [CrossRef]
40. Lianjun, W.Y.; Yabian, Y.Q. Screening aerobic denitrifiers from soil and study on characteristics of denitrification. *Chin. J. Environ. Eng.* **2011**, *5*, 1902–1906.
41. Zuo, W.; Harbin, D. An Aerobic Denitrifier Screened Identification and Denitrification Characteristic. Master's Thesis, Harbin Institute of Technology, Harbin, China, 2006.
42. He, T.; Xu, Y.; Li, Z. Identification and characterization of a hypothermia nitrite bacterium *Pseudomonas tolaasii* Y-11. *Acta Microbiol. Sin.* **2018**, *55*, 992–1000.
43. Lee, D.U.; Lee, I.S.; Choi, Y.D.; Bae, J.H. Effects of external carbon source and empty bed contact time on simultaneous heterotrophic and sulfur-utilizing autotrophic denitrification. *Process Biochem.* **2001**, *36*, 1215–1224. [CrossRef]
44. Huang, X.F.; Li, W.G.; Zhang, D.Y.; Qin, W. Ammonium removal by a novel oligotrophic *Acinetobacter* sp. Y16 capable of heterotrophic nitrification aerobic denitrification at low temperature. *Bioresour. Technol.* **2013**, *146*, 44–50. [CrossRef] [PubMed]
45. Song, Y.J.; Li, Y.; Liu, Y.X.; He, W.L. Effect of carbon and nitrogen sources on nitrogen removal by a heterotrophic nitrification-aerobic denitrification strain Y1. *Acta Sci. Circum.* **2013**, *33*, 2491–2497.
46. Duan, J.M.; Fang, H.D.; Su, B.; Chen, J.F.; Lin, J.M. Characterization of a halophilic heterotrophic nitrification-aerobic denitrification bacterium and its application on treatment of saline wastewater. *Bioresour. Technol.* **2015**, *179*, 421–428. [CrossRef]
47. Ren, Y.X.; Yang, L.; Liang, X. The characteristics of a novel heterotrophic nitrifying and aerobic denitrifying bacterium, *Acinetobacter junii* YB. *Bioresour. Technol.* **2014**, *71*, 1–9. [CrossRef] [PubMed]
48. Zhang, Q.L.; Liu, Y.; Ai, G.M.; Miao, L.L.; Zheng, H.Y.; Liu, Z.P. The characteristics of a novel heterotrophic nitrification–aerobic denitrification bacterium, *Bacillus methylotrophicus* strain L7. *Bioresour. Technol.* **2012**, *108*, 35–44. [CrossRef] [PubMed]
49. Kim, D.J.; Lee, D.I.; Keller, J. Effect of temperature and free ammonia on nitrification and nitrite accumulation in landfill leachate and analysis of its nitrifying bacterial community by FISH. *Bioresour. Technol.* **2006**, *97*, 459–468. [CrossRef]
50. Rolf, A. Inhibition of *Staphylococcus aureus* and spheroplasts of *Gram-negative bacteria* by an antagonistic compound produced by a strain of *Lactobacillus plantarum*. *Int. J. Food Microbiol.* **1986**, *3*, 149–160.

www.ingramcontent.com/pod-product-compliance
Lightning Source LLC
Chambersburg PA
CBHW062105220926
51850CB00010B/2904

MDPI
St. Alban-Anlage 66
4052 Basel
Switzerland
Tel. +41 61 683 77 34
Fax +41 61 302 89 18
www.mdpi.com

Water Editorial Office
E-mail: water@mdpi.com
www.mdpi.com/journal/water